Concrete pumping and spraying

A practical guide

About this book

This informative and comprehensive book demonstrates the commercial viability of concrete pumps and sprays, and explains the practical advantages of concrete pumping and spraying over traditional methods of concrete placing. It is a practical guide giving advice on corrective measures for common problems and guidance on safety and training, together with hints for concrete pump operators. This information is invaluable to those involved in concrete pumping and spraying, and all those considering the adoption of this method of concrete placing.

The demand for efficient and rapid concrete placing methods has resulted in the development and refinement of concrete pumping plant and equipment, and its use is becoming commonplace throughout the construction industry. Reductions in contract time which can be achieved by using concrete pumps may result in overall savings and higher competitive margins for construction companies.

Tom Cooke has spent most of his working life in civil engineering contracting. In the immediate post war years he was involved in heavy civil engineering projects, and was particularly involved with all types of piling and foundations. He was invited to join the GKN organization in 1963 in a newly formed piling and ground engineering company as Regional Manager, Scotland. Later he was Contracts Director, Piling Division of GKN Keller Foundations, Coventry.

His interest in concrete pumping was stimulated by his research into this subject in recent years, in association with the University of Lancaster. He is well known as consultant to the British Concrete Pumping Association, has written many articles on the subject and lectured extensively throughout the country.

Concrete pumping and spraying

A practical guide

T. H. Cooke

Thomas Telford, London

Published by Thomas Telford Ltd, Thomas Telford House, 1 Heron Quay, London E14 9XF

First published 1990

Cover illustration courtesy F.W. Schwing/Burlington Engineers

British Library Cataloguing in Publication Data
Cooke, Tom
Concrete pumping and spraying: a practical guide
1. Sprayed concrete
I. Title
666'.893

ISBN: 0 7277 1505 4

Typeset in Great Britain by the Alden Press
Printed and bound in Great Britain by Redwood Burn Limited, Trowbridge, Wilts

Acknowledgements

The author is grateful to the following individuals, firms and organizations who kindly offered assistance in the preparation of this book.

Audley Ackermann; Aliva Ltd; Allentown Pneumatic Gun Inc.; American Concrete Pumping Association; American Pecco Corporation; Ash Marketing, Central Electricity Generating Board; Brian Bennett; Andrew Boath; British Cement Association; British Concrete Pumping Association Ltd; Burlington Engineers Ltd; Cement Admixtures Association; CIFA SpA; Brian Doig; Tony Dwyer; Elba-Werk GmbH; Elkem Chemicals Ltd; John Fairley; Allan Fawcett; Peter Garner; Geotextiles Projects Ltd; GKN Keller Foundations; Ian Gooding; Greenhill Plant; Jim Henderson; James R. Hubbard; Hydroseal Concrete Pumps Ltd; John Kluger; Wolfgang Kowalski; Department of Engineering, Lancaster University; Trevor Lupton; Richard Majec; Roger Manning; Tony Meares; Mokan Ltd; David Mundell; Medeo Olivares; Pipejoint & Co.; Michael Pochin; Pochin Group plc; Premier Concrete Pumping Ltd; Putzmeister GmbH; Putzmeister Ltd; Rapid Concrete Placing Ltd; William C. Roberts; Royston Conpump Ltd; Peter Schlickmann; F.W. Schwing GmbH; Shepherd Construction Ltd; Shotcrete Ltd; Sidewinder Concrete Pumps Ltd; Sprayed Concrete Association; Sprayed Concrete Ltd; Steetley Construction Materials Ltd; David Terrill; Transcrete (UK) Ltd; Utranazz (Bourne End) Ltd; Jack Walker; Paul Wells; Henley Wheadon; Eric Windsor; Windsor Concrete Pumping Ltd.

Special thanks are extended to John Illingworth for his ever sound advice and encouragement and to Dr Mike Anson (formerly head of Department of Engineering, University of Lancaster, now head of Department of Civil and Structural Engineering, Hong Kong Polytechnic), with whom the author shared many months of interesting concrete pumping research.

Contents

1

Introduction

The pumping of concrete is universally accepted as one of the main methods of concrete distribution and placement. Although concrete pumping may not always be the most favourable placing option, it offers certain advantages over other methods such as crane and skip, hoists and conveyors. The use of concrete pumps is increasing throughout the world as the pumps become more reliable. The concrete mixes that enable concrete to be pumped are also understood much better.

1.1. Development of the concrete pump

The first use of a pump for placing concrete is not recorded, but the first patent for a concrete pump was taken out in the USA in 1913. In about 1930 the German firm Torkret GmbH, started producing a single cylinder (180 mm dia.) mortar pump. In the 1930s research and development continued and pumps were built under licence in several countries including the USA, France, Holland and the UK; however, much design improvement was still required. The sliding plate valve was introduced during this period and superseded the early ball valves.

When rebuilding started in Europe after the Second World War, pumps were still mechanically driven. The next decade saw the introduction of the hydraulic drive for concrete pumps. In West Germany in the 1950s and 1960s, the *Wirtschaft Wunder* (economic miracle) gave an impetus to the development of the concrete pump. The West Germans were rebuilding rapidly and concrete pumping became widely used. As a result the West Germans are world leaders in the manufacture of concrete pumps and place nearly 40% of their concrete by pump, compared with 10% in the UK. A report in 1986[1] comparing UK and West German practice, noted the difference in attitude towards concrete pumping. The

Fig. 1. Mobile concrete pump with 52 m boom (courtesy F. W. Schwing/Burlington Engineers)

West Germans used the pump simply 'because it was quicker', whereas in the UK there was still reluctance to use a method which seemed more costly than the traditional crane and skip option. Immediately after the Second World War, pump users in the UK had problems and there were reservations about the reliability of the pumps. The users blamed the pumps for causing blockages which more probably were caused by the concrete mix. However, advances in mix design for pumping and improved valve design have resulted in a steady increase in the use of pumps in the UK. Since the end of the Second World War contractors in many countries have seen that this method of placing concrete is the most appropriate for many of their pours. After deciding to use concrete pumps, the contractors automatically altered their working practices in, for example, design (e.g. slab sizes), formwork, steel-fixing, and concrete supply to facilitate pumping. In other words, concrete pumping was always considered and was an integral part of programming and organizing a project. In some countries, however, the tendency is still to use a pump if placing cannot be carried out readily by any other method.

Even on the small construction site in West Germany, where tower cranes are ubiquitous, the day to day small pours are carried out by concrete pump. The site foreman's answer to the question, 'Why are you using a pump when you have a crane on site?' is invariably 'Because it's quicker and in any case if the crane is used for concreting, other work will suffer'.

The keen rivalry between the leading German manufacturers, Schwing, Putzmeister and Elba-Scheele, has boosted the development of the concrete pump and in particular the valve design. From time to time world records for the length, height and placing rates of a concrete pour are broken which shows the manufacturers' desire to improve efficiency and performance. This can only be a good thing for the construction industry world-wide.

1.2. The need for concrete pumping

The use of a concrete pump is one means of speeding up construction. That alone is surely a good indication of why there is a need for concrete pumping.

The financial effect of speed of construction is clearly important. The client requires the earliest possible return on investment and the contractor requires the highest return on capital employed. The necessity for concrete pumping on major civil engineering projects where large quantities of concrete are required, both in total and in individual pours, is universally recognized. Without this placing option the construction time on such work would no doubt be unacceptable, for example in offshore oil platform construction or power stations. The concrete pump is also essential in most tunnel construction, where concrete may have to be conveyed over long distances in restricted access. Alternatively, a tunnel train consisting of a materials hopper, mixing unit and concrete pump, can be placed inside the tunnel and moved forward as the work progresses (see chapter 5). Advances in concrete technology have facilitated the continuous casting of large pours for which the concrete pump has been recommended.[2] For such pours this placing method can be said to be technically the most appropriate as well as the quickest.

The programmed use of a concrete pump can be beneficial in fast-track construction. But the fast-track philosophy can only be fully exploited by contractors if designers design and specify for speed. Erection time may be considered to be a factor in design. Pour sizes, for example, could be increased allowing scope for the use of a pump. There could also be significant savings in stop-end costs if larger pours are used. The opportunity of pouring a large slab in one pour instead of several smaller pours must be welcomed by contractors.

Concrete pumps could be used to better effect in multi-storey building construction. Concrete placing rates are maintained whatever the height, which is not the case when using a crane or hoist. Most high-rise sites in urban areas are congested and conveying concrete through a pipeline is more efficient than using dump trucks, hoists or cranes. Crane work-loads can certainly be reduced if a pump is scheduled to place the concrete.

There have been many construction projects when lost time has been retrieved by reprogramming around the use of a concrete pump which will place the material at over twice the speed of placing by crane and skip.

A concrete pump which provides a method of placing concrete more efficiently and more quickly to greater heights and lengths than other methods is an item of equipment which should be fully utilized by the construction industry.

1.3. Advantages of concrete pumping

Using a pump offers certain advantages over other methods of concrete placement. Transporting concrete in a steady flow from the delivery point directly to the pour position removes the need for long barrow runs, power buggies, staging, hoists, dump trucks, etc., or a combination of these items.

Thus the concrete pump provides flexibility in overcoming access problems, which is a distinct advantage over other placing options. The pour may be outside the range of the crane or narrow delivery routes may preclude the use of truckmixers, dump trucks or even barrows. Figs 2 and 3 show how an access problem was overcome when concrete had to be moved 200 m up a hillside to the pour position at Edale, Derbyshire. It was not possible for ready-mixed concrete trucks to travel up to the pour position. At the pour position, the concrete had to be placed in a confined space between narrow formwork. The concrete pump coped with the problem easily.

Cost considerations are of prime importance. Apart from direct discharge from a ready-mixed concrete truck, placing by pump is the quickest method. Reduced costs can be significant particularly with regard to overall programme time savings. It is important, however, that proper methods are used to assess such costs (see chapter 6).

When high volume pours are required to be completed within a stipulated period of time, the concrete pump is very

Fig. 2. Pumping uphill through a static line to pour position (courtesy Pochin Group plc)

often chosen because other methods would be too slow. Several high output pumps are sometimes used on the same pour to meet production requirements. The concrete supply must obviously keep the pumps operating at full load. A good example of high volume work was the placing of 3000 m^3 of concrete in Staffordshire in 24 hours' continuous pumping. Two large mobile pumps were used.

Placing rates can be maintained by a pump when placing at heights, but this is not possible when using other methods. For example, when using a crane and skip the higher the pour the longer it will take to raise and lower the skip. In this context it is worthwhile to compare typical concrete placing

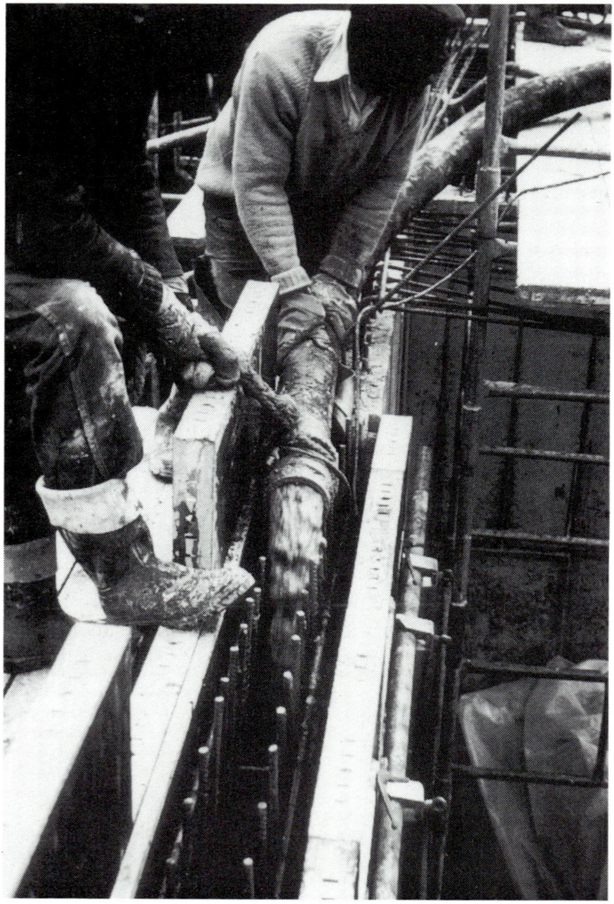

Fig. 3. Pumping concrete into a narrow gap between formwork (courtesy Pochin Group plc)

performances of the concrete pump with the crane/skip option. This reveals distinct advantages in placing rates using the pump. On the Frankfurt Post Office Tower project in 1977 when concrete was pumped to the then world record height of 310 m, the saving in concreting time using a pump instead of a crane and skip was estimated to be 40%. This figure cannot be established accurately, but it can be considered to be a realistic or reasonable estimate.

Estimating concreting speeds or outputs using a crane requires information on the lifting height which can vary from a few feet to hundreds of feet and the slewing circle could be from 0° to 360°. At best the output from crane/skip placing will rarely exceed 12 m^3/h. The average output is well below this, certainly no more than 10 m^3/h. The figure generally assumed for a satisfactory average output from a concrete pump is 25 m^3/h.

Assuming that a truckmixer has backed up to a concrete pump and is ready to discharge the concrete into the pump hopper, the pump can accept a full 6 m^3 load of concrete and pump it in five or six minutes under normal conditions. The time for a crane and skip to place the same quantity of concrete would depend very much on the circumstances under which the crane is operating. In addition, the size of the crane and the radius at which it is operating will determine the skip capacity which can be used. Having to use a small skip will significantly prolong the placing time: the speed of placing will be directly proportional to the size of the skip.

When using a crane in the UK, the Construction (Lifting Operations) Regulations[3] and CP3010[4] must be complied with. Other countries have similar regulations. The weather can influence the operation of tower cranes particularly, as certain wind conditions will inhibit their use.

With such constraints it is difficult to estimate an average time it takes to discharge a 6 m^3 truckmixer when using a tower crane. Depending on conditions, this would be at least 15 minutes. A reasonable figure would be between 15 and 25 minutes, say 20 minutes. Therefore comparing the time taken to empty a 6 m^3 truckmixer when placing concrete by pump and by crane, of, say, 6 minutes, and say 20 minutes, respectively, it can be seen that emptying the truckmixer when using a pump is about three times quicker than when using a crane. However, this does not mean that the pour will be about three times quicker if a concrete pump is used. To achieve this there would have to be an uninterrupted concrete supply with space for two truckmixers to position themselves at the pump hopper so that pumping could be continuous. The crane/skip operation is not so adversely affected in the same way, as the next truckmixer has time to manoeuvre into position while the skip is being raised, discharged and lowered again. In practice the ratio is usually about 1:2·5; it was mentioned earlier that the average outputs for the crane/skip and the concrete pump are 10 m^3/h and 25 m^3/h, respectively.

Crane/skip pours can be speeded up by ensuring that the largest skip possible is used and that there are two of them available, i.e. one is being filled as the other is being discharged. If it is possible to use a crawler crane instead of a tower crane, the pour will be much faster.

Even with these advantages which will increase the placing rate of the crane/skip method, the output from the concrete pump will always be substantially greater, if the standard of concrete supply service, efficiency of placing crew, etc. are the same. The potential offered by a concrete pump which will do the job 2–2·5 times faster is unquestionable. To go further and attempt to quote comparative costs or hire figures of concrete pumps, tower cranes, etc., is beyond the scope of this book as rates vary so much throughout individual countries and the world. In addition, each contract is different and must always be treated as such.

Labour placing costs are generally less using a pump as on average the gang size will be smaller (usually by one or two people, depending on the type of pour), principally because distribution with the flexible end hose or distribution arm is much easier. In addition, a reduction in the total pour placing time also reduces the labour costs for the operation.

The quality of the concrete is also maintained in adverse weather conditions, as the material is protected in the pipeline, irrespective of the length it has to be transported.

The concrete pump is one of several items of dominant plant on the construction site around which the work has to be planned. Although this subject is primarily concerned with planning, it is included here to show that if the site management recognizes that the pump is an item of dominant plant, the management performance can be improved, resources can be used better and production can be increased. Illingworth[5] discusses this important aspect of planning in *Site handling equipment.* Inefficient use of dominant plant such as a tower crane can have an extremely adverse effect on the site. The tower crane may be fully occupied carrying out jobs such as handling formwork, steelwork, staging, scaffolding, small plant items and so on. It is possible that simultaneously a concrete pour may be proceeding and that the crane is required to handle the concrete skips. The result in this case could be disastrous, depending on how much lifting work is going on concurrently. For optimum concrete placing rates it is clear that when using the crane/skip option, the crane should be solely occupied by placing concrete, but this seldom occurs. The result is generally a series of delays with every operation suffering. Demand for attention by individuals easily leads to friction, frustration and to loss of production. It is a matter of optimum distribution of hook time. If this is miscalculated at the tender stage, the effect during construction will become apparent: underestimating will result in delays and overestimating will result in the crane not being used fully.

Recognizing dominant plant and using it efficiently requires careful planning and programming of the site construction sequences or building cycles such as in multi-storey work. The use of the tower crane and the various handling demands made on it have to be carefully controlled. Discipline is essential. If the tower crane is used efficiently, certain advantages will become apparent. For example, the management becomes aware of the need for proper planning in the use of dominant plant and this in turn creates better management. Furthermore, idle dominant plant is very noticeable and calls for immediate action to utilize the plant properly, which again improves the management.

As the concrete pump is another item of dominant plant it is therefore a key unit on site. Its efficient use depends on how well the planning has been carried out around its operation. To obtain the best performance out of a concrete pump (as the dominant plant) the actual pumping time must be as short as possible. To obtain this there are four main points which require attention

(*a*) the pour must be ready on time
(*b*) there must be a sufficient number of experienced people in the placing crew
(*c*) the concrete supply must be available at the start and throughout the pour
(*d*) the concrete must be pumpable.

All this looks straightforward, yet inadequate planning of any one of these items will delay the whole operation.

The time saving on the whole of the concreting operations which cover the contract — preparation, mix, haul and place — may not appear to be significant at the time when any one pour takes place, but over the whole of the contract the savings in time, and consequently in money, could be substantial.

What are the benefits which may be evident if in fact the planning and execution of the pour is good? First, from (*a*) there might be improved performance from formwork and steel-fixing gangs having to meet a deadline, (*b*) places the onus on site management to see that the placing gang is properly trained and experienced; (*c*) and (*d*) are a timely reminder of the close co-operation and planning necessary between concrete supplier and contractor when the speed of the pour and concrete specification (with pumpable concrete in mind) are discussed well in advance of the pour. These constraints without doubt can be beneficial if they are dealt with effectively.

Thus if the role of dominant plant is recognized by site management, this can lead to improved management performance, better use of resources and increased production.

1.4. Concrete pumping industry world-wide

The concrete pumping market in Japan is said to be the largest in the world. Since 1988 Japan has produced roughly the same number of pumps as West Germany, but currently only exports about 5%. It is estimated that there are 300 pumps operating in Tokyo alone. Half of the pumps in Japan are small mobile units.

The second biggest market for concrete pumps is the USA. Most pumps manufactured in the USA are produced under license from the West German manufacturers Schwing and Putzmeister; the number of local manufacturers is relatively small.

These two leading West German manufacturers also supply the majority of pumps used in most other countries throughout the world. West Germany exports about 70% of its concrete pumps. An exception, however, is the Italian firm CIFA which is the largest in Italy and supplies the bulk of the concrete pump requirements in that country and also exports concrete pumps.

Number of pumps in use

It is virtually impossible to give the total number of concrete pumps operating throughout the world. Few countries produce statistics on the number of pumps operating, perhaps simply because there is no particular need to have this information. The manufacturing companies will have sales records, but what is not known is the number of pumps operating, say, in any one year. The number of pumps which are damaged beyond repair or written off and taken out of service cannot be estimated to any degree of accuracy.

As might be expected, the number of concrete pumps operating in West Germany exceeds that of other European countries and official road registration figures show that there are about 2000 mobile units. There may be a similar number of portable/trailer pumps, but there are no published figures. In the UK there are relatively few pumps (about 400) compared with other countries such as France, where it is estimated there are 1000 mobile units. The important thing, however, is that the numbers are growing steadily and will no doubt continue to do so. In the USA there is again no formal record of the number of pumps, but the ACPA say that a reasonable estimate would be that there are over 7000 pumps of all types.

Percentage of concrete pumped in various countries

It is interesting to try and ascertain a comparison of the percentage of total site placed concrete which is pumped in different countries. The only records of this are to be found in the ERMCO records of recent years. The figures refer to ERMCO associated companies only, but they can be considered to be a reasonable guide. Using these figures and making an allowance for site batched concrete which is pumped (ERMCO statistics concern ready-mixed concrete only), the percentage of total site placed concrete which is placed by pump can be calculated roughly (Table 1). It is not known why the Greeks pump the highest percentage of site-placed concrete. The average percentage for ERMCO countries is probably 20–25%, with the UK, Spain and Ireland well down the list. West Germany places nearly twice as much concrete by pump on site as the UK despite having a similar population, area, size of construction industry, etc. Perhaps one reason for some of the difference is that nearly every house in West Germany has a basement: the average house basement contains over 30 m^3 of concrete. Most of the concrete for these basements is placed by pump (Fig. 4). The Japanese pump 30–40% of ready-mixed concrete supplies (the aggregate content is mainly crushed material), whereas in the USA the percentage pumped of total site placed concrete varies from as low as 5% in certain areas to as high as 45%.

*Table 1. Percentage of total site placed concrete which is pumped**

Country	% placed by pump
Austria	18
Belgium	20
Finland	20
France	28
West Germany	35
UK	10
Greece	85
Ireland	5
Israel	40
Italy	28
Netherlands	15
Norway	19
Spain	12
Sweden	28
Switzerland	15

*ERMCO figures for 1984, for associated companies ready-mixed supply. Allowance added to ERMCO figures for site batched concrete estimated to be pumped.

Pump hire industries

In Japan, Australia and most other countries the pump hire companies are small, the majority owning perhaps only one or two pumps. The ACPA has approximately 200 members who own pumps: the ownership is split roughly as given in

Fig. 4. Concreting the basement of a house in West Germany

Table 2. This follows the ownership pattern in West Germany, where the majority of the hire companies own only a few pumps. However, unlike in the USA some West German ready-mixed concrete companies also own concrete pumps, or at least have a close association with the pump hirers. In the USA it is felt that commercially it is unfavourable for ready-mixed concrete companies to own pumps, as contractors would insist on the companies supplying the pump free of charge to gain the concrete supply order.

The Lancaster University Report[1] compared the pump hire industries in the UK and West Germany. In both countries approximately 20% of pumps are owned by contractors. The remaining 80% in West Germany is split between ready-mixed concrete companies (50%) and pump hire companies (30%). In the UK, ready-mixed concrete companies used to own most of the pumps, but this is no longer the case. The remaining 80% in the UK today are owned and operated by about 30 pump hire companies.

The pump hire companies in West Germany tend to be much smaller than in the UK and average only two or three pumps per company, but in the UK several companies own

Table 2. Breakdown of ownership of concrete pumps in the USA

Number of pumps owned	Percentage
1	36
2–4	32
5–9	19
10 or more	13
Total	100

more than twelve pumps. One such company, the Pochin Group plc, with headquarters in Cheshire, owns over 100 units and is by far the leader and largest concrete pump hire company in Europe and probably in the world. This company also sets the highest standards in mechanical efficiency and general appearance of its plant and has done a great deal to project the image of concrete pumping in the UK favourably. Visitors from other countries cannot fail to be impressed. The company has introduced mandatory procedures for operator training, plant maintenance and safety, which are rigidly adhered to. Pochin can be said to lead by example and many pump hire firms would produce better results and performances by insisting on formalized training and written procedures. The UK concrete pump hire industry's continuing growth and favourable image has come from the example of this large company.

2

Concrete pumping plant and equipment

The modern concrete pump is a sophisticated, reliable and robust machine. It is produced in many shapes and sizes to suit the numerous applications and demands of the construction site. Equally important is the distribution equipment required to transport the concrete from pump to placing position.

2.1. Types of concrete pump and valve systems

A great deal of development has taken place since the early days of concrete pumping when the pumps were mechanically driven, incorporating single or double pistons. The simple two-stroke mechanical pump consisted of a receiving hopper, an inlet and an outlet valve, a piston and a cylinder. The pump was powered by a diesel engine. The pumping action starts with the suction stroke (assisted by gravity) drawing concrete into the cylinder as the piston moves backwards. During this operation the outlet valve is closed. On the forward (pumping) stroke the inlet valve closes and the outlet valve opens to allow concrete to be pushed into the delivery pipeline (Fig. 5). The modern concrete pump still operates on the same principles but develop-

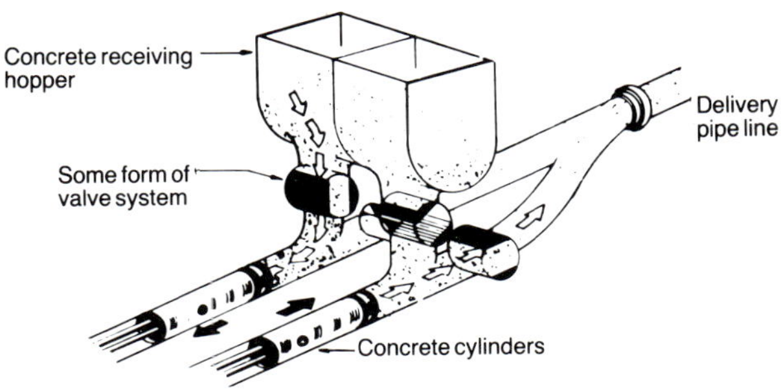

Fig. 5. Principle of the concrete pump

ment has produced vast improvements and refinements resulting in concrete pumps being reliable and invaluable pieces of equipment for the construction industry in all parts of the world.

Pneumatic placers are another early form of concrete pump, consisting of a compressed air chamber which, when filled with concrete, is sealed and compressed air added. This pressurizes the concrete to move it through the pipeline. Early models created problems with the dangerously high pressures involved, but development resulted in lower pressures and the units are now much safer to operate. There are relatively few still in use, mainly because of disadvantages in height and distance pumped and because higher slump concrete is required. However, they are used extensively for conveying mortar screeds (see chapter 10).

In 1963, in the USA, the peristaltic or squeeze type pump was developed. The principle is illustrated in Fig. 6. Concrete is forced through a rubber tube by rollers. The rubber tube is located within a chamber where the pressure is lower than atmospheric pressure so that the tube can regain its shape after the rollers have passed over it. This provides suction to assist concrete flow from the hopper to the pipeline. The West German firm Wibau manufactured many of these pumps, claiming advantages in smooth concrete flow and ability to pump at minimum pressure.

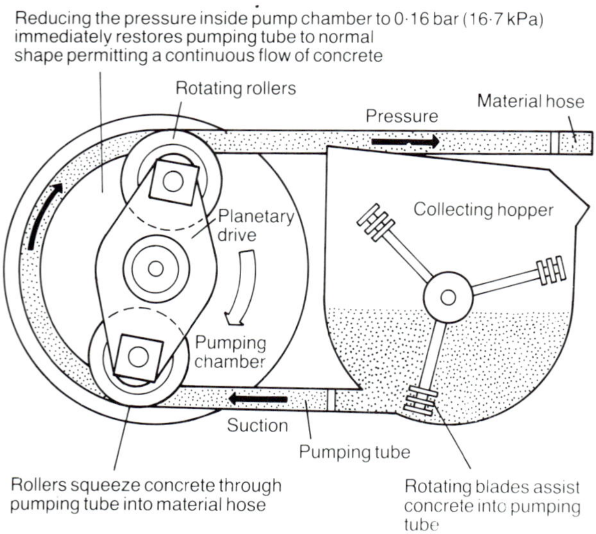

Fig. 6. Principle of the peristaltic pump

The hydraulic piston pump is the most widely used modern concrete pump. There is a wide range available and the manufacturers claim that there is a pump suitable for any application. Specifications differ, but the concept of a concrete pump is still the same as it was for the original mechanically driven pumps. A pump consists of three parts: a concrete receiving hopper, a valve system and a power transmission system. Designs vary considerably depending on the type of pump and on the purpose for which it was designed. The customer has a very wide choice of pumps.

There are three main types of concrete pump: mobile, trailer or static and screed or mortar. The pumps are mounted on some form of carrier. A mobile pump will have a placing boom. Independent placing booms are used with both mobile and trailer pumps. These pumps are described in detail later, but it is appropriate first to describe the main valve systems which are used.

Types of valve

The most important part of any concrete pump is the valve system and several types have been developed over the years. Most of the problems in concrete pumping have concerned valve design and this is why the manufacturers have spent so much time and money in looking for improvements. The main types of valve are: peristaltic or squeeze type valves; sliding gate or rotating visor valves; flapper valves; and hollow transfer tube valves.

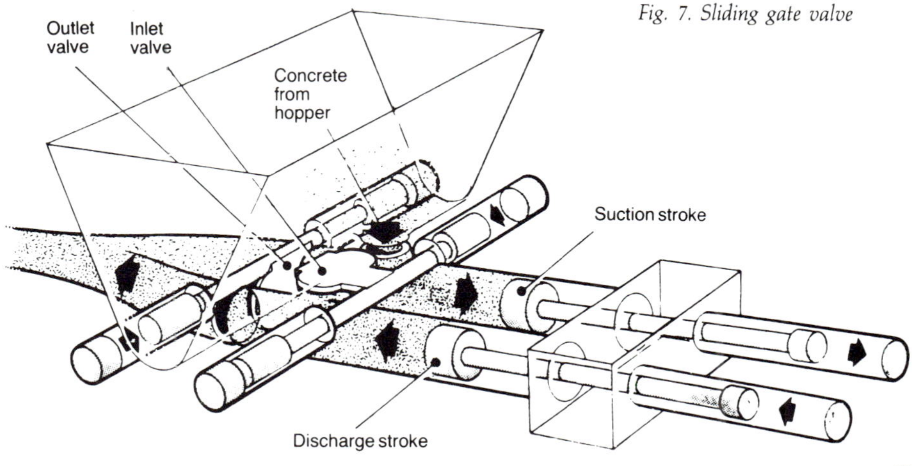

Fig. 7. Sliding gate valve

Sliding gate valves or rotating visor valves are well proven and can be very effective. Concrete from the two cylinders is brought together by means of a Y pipe or so-called Siamese pipe (Fig. 7). This type of valve system was one of the earliest to be used. It is modified and improved from time to time and is still widely used.

Flapper valves are also widely used for straightforward applications, as they are simple and relatively cheap. Some pressure loss may occur and difficulties may arise with certain mixes. However, it should be remembered that there is only one valve and although it has poor wearing qualities it can easily be reconditioned. These valves have been used for many years which is proof of their usefulness. Fig. 8 shows a typical flapper valve arrangement.

Hollow transfer tube valves are the most commonly used type of valve. One type which is used extensively is the Rock valve (Fig. 9). This valve offers a balanced changeover under any condition with a high sealing efficiency. The cutting ring maintains contact with the wearing plate at all times. The two main wearing parts give minimal wear and parts can be replaced easily and speedily: there is a pivoting pipe outlet which gives internal access. It is claimed that

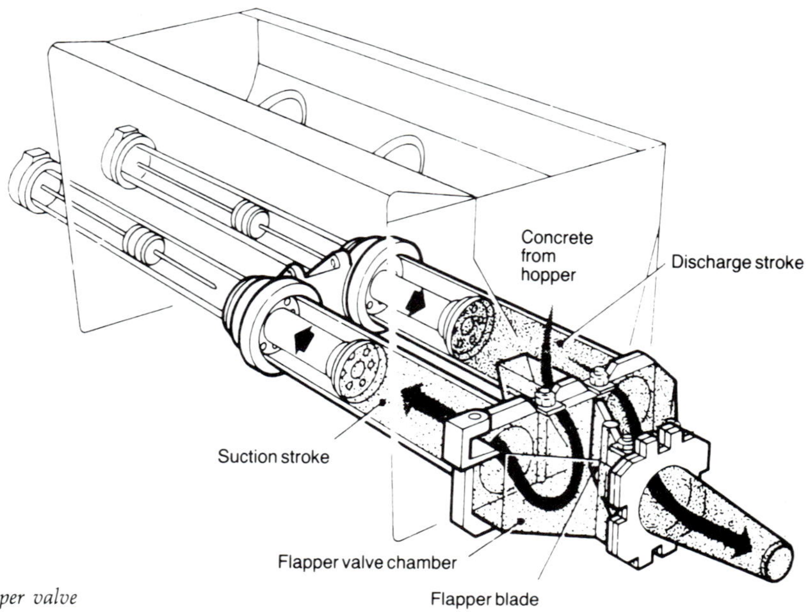

Fig. 8. Flapper valve

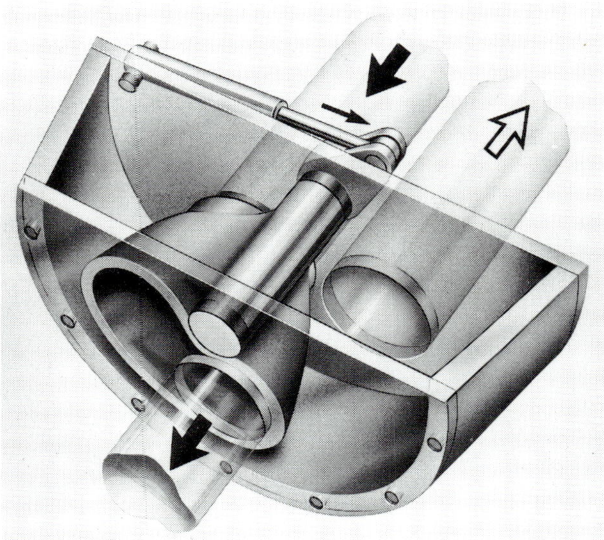

Fig. 9. Rock valve

there are few problems with cleaning or maintenance. Schwing asserts that the Rock valve will cope with extreme concrete mixes and this has been proved in practice.

The S valve used by Putzmeister is another example of a transfer tube valve; the reduction S shaped pipe is located inside the hopper. It is claimed that the parabolic reduction and specially designed automatic sealing ring can deal with the most difficult of concrete mixes and grout leakage is eliminated. The inertia forces of the S-transfer tube are small despite speedy transfer movement of the tube. Fig. 10 shows the S valve with a 400 l hopper.

The elephant trunk system, which was first introduced in 1972, is also commonly used and offers several advantages. As it avoids a complex valve arrangement or Y pipes, the cylinders open directly into the hopper. There is little loss of pressure during cylinder discharge and little power is required to operate the swinging pipe (Fig. 11). The C shaped steel trunk swings to and fro in front of the delivery cylinder openings. This system is extremely simple yet effective. The automatic ring which is located at the only junction in the elephant system is a simple sealing and wear element. The ring always presses automatically against the cylinder openings and the manufacturers claim the improved sealing reduces bleeding even with mixes lacking in fines. The trunk is made of high quality wear-resistant steel, has a

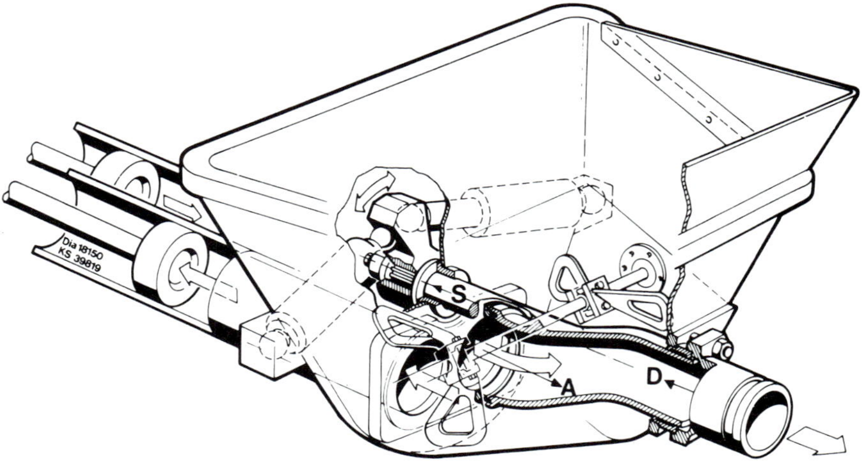

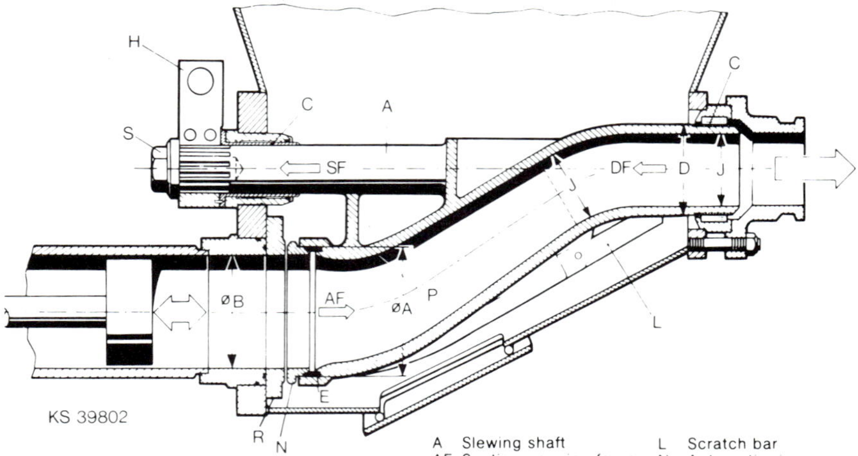

A Slewing shaft
AF Suction opening force
B Suction dia.
D Eff. pressure tube
DF Pressure outlet force
E Rubber seal
H Switch lever
J Outlet inner dia. 125 mm
L Scratch bar
N Automatic ring
P Parabolic reduced dia.
R Spectacle plate
S Adjustment nut
SF Axial force
C Chrome bushing

Fig. 10. S valve with 400 l hopper

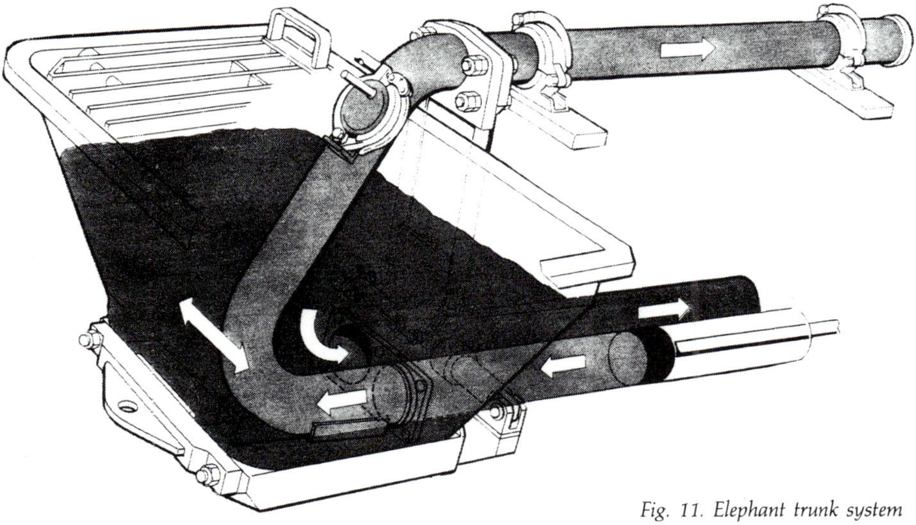

Fig. 11. Elephant trunk system

wall thickness of up to 20 mm and will handle over 50 000 m^3 before replacement is necessary.

Two UK concrete pump manufacturers have developed their own valve arrangement. The A valve, developed by Mokan Ltd, is a simple swinging pipe arrangement offering good flow characteristics and complete agitation of the concrete in the hopper. Wear is reduced by sharing the concrete flow between the two arms of the valve and

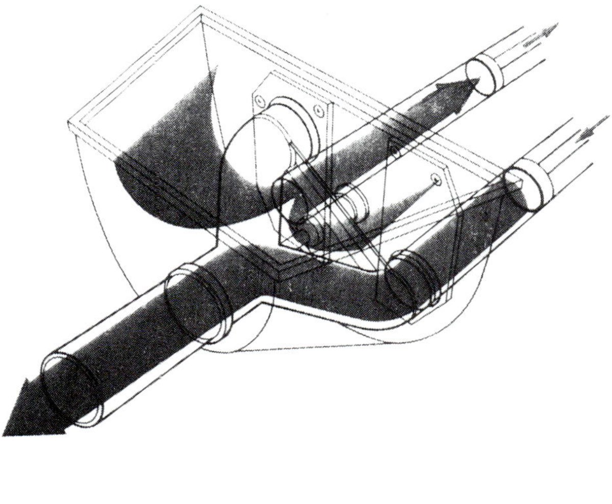

Fig. 12. Mokan A valve

servicing adjustments can be carried out speedily. This valve arrangement is shown in Fig. 12.

The Pulser valve is another relatively new transfer tube type of valve used in Hydroseal pumps. The valve acts on the transfer tube shaft and pulls the face of the transfer tube tightly against the concrete cylinder outlet, giving automatic sealing. It is claimed that the clamping pressure is maintained at 15% greater than that generated by the concrete pressure, thus minimizing grout leakage.

These various valve systems form an integral part of the different types of concrete pump. Each manufacturer will claim advantages over the others with a particular valve arrangement, but in general it can be said that the systems used nowadays are generally simpler, have more positive sealing, are more wear resistant and can be maintained easily. The main types of concrete pumps will now be examined in more detail.

Mobile pumps

Mobile pumps were first developed by Schwing in 1965 and consist of a concrete pump unit mounted on a vehicle chassis, which may have two, four, six or even eight axles depending on the size of the unit. The folding distribution booms incorporated in the design vary in length from 15 m to over 50 m and are constructed in 2–5 sections which can be folded up when travelling (Fig. 13). Thus there is a full

Fig. 13. Typical mobile pump with 38 m boom (courtesy Pochin Group plc)

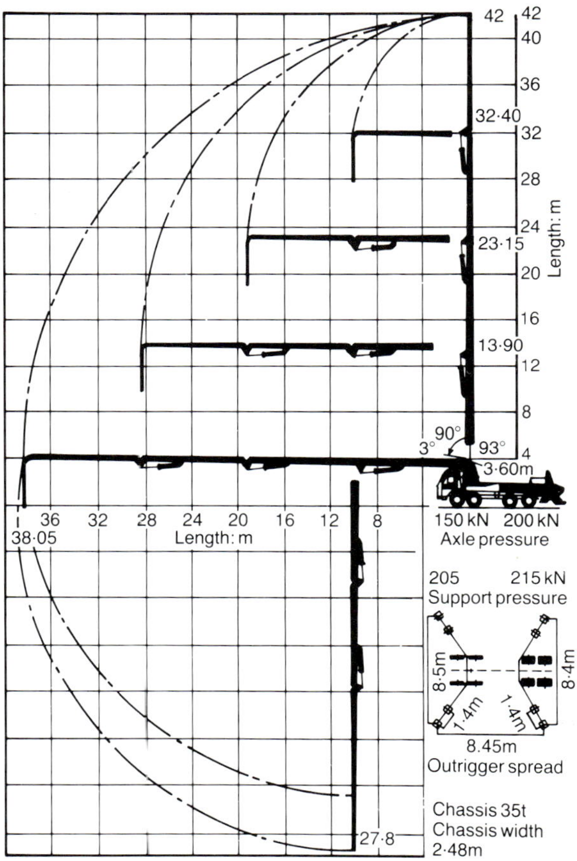

Fig. 14. Working range of KMV42 mobile pump (courtesy F. W. Schwing/Burlington Engineers)

range of mobile pumps available to cover most situations and demands of the construction industry. These pumps are fully hydraulic and are extremely versatile, principally because of their mobility and adaptability. It is not uncommon for mobile pumps to work on several sites on the same day, sometimes as many as five or six. This requires first class communication and an uninterrupted concrete supply. The full circle slewing affords distinct advantages on confined sites with restricted work space, enabling concrete to be placed exactly where it is required. Most concrete pump manufacturers provide diagrams showing the working range of the boom (Fig. 14). Mobile pumps can be used without the boom if long or excessive lengths of pipe are required.

Mobile pumps now have electrically (by cable) or radio

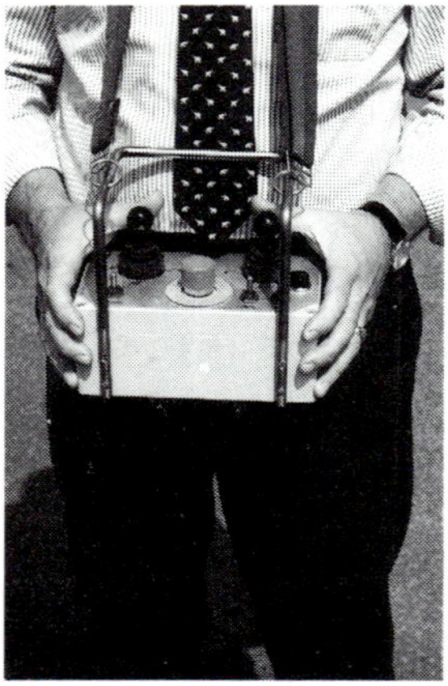

Fig. 15. Remote control box

Fig. 16. Pump operator's view on the Nord-Ostsee Kanal, Brünsbuttel, West Germany

operated remote control facilities enabling the operator to control the pumping operation from the most advantageous position, preferably observing the hose end (Figs 15 and 16).

Special remote controlled equipment was installed in mobile pumps used at Chernobyl (see chapter 3), together with video equipment, enabling the pump to be operated from within a lead shielded cab remote from the radiation contaminated area. The operator observed all functions on a monitor from a camera mounted on the boom arm nearest the placing hose and another attached to the outrigger leg covering operations such as filling the hopper, and monitoring pressure gauges (Fig. 17). A further innovation is the development of an automatic mast control (AMC) which enables all boom movements to be controlled remotely by a single lever.

Concrete pumps and particularly the mobile pumps with boom attachments are now both sophisticated and practical. The normally high reserves in power capacities enable the pumps to work well within their capabilities: this is important for trouble-free pumping.

Fig. 17. Video control cameras on boom and outrigger leg (courtesy Putzmeister Ltd)

Portable concrete pumps

Portable concrete pumps are compact and robust units and are produced in a wide range of sizes and for an equally wide range of operations. Outputs range from the small multi-purpose pump rated at about 30 m^3/h to the large stationary units with a theoretical rating of 135 m^3/h.

These machines are sometimes called trailer, stationary or static concrete pumps. Trailer pumps, as the name suggests, can be towed from site to site as required and are mainly mounted on a single-axle chassis. The larger stationary/static pumps are again mounted on a single-axle chassis and can also be towed. The trailer pump is used for smaller pours, perhaps on short-term hire arrangements, whereas stationary pumps (as the name implies) are usually to be found on large contracts at fixed positions, distributing concrete to various parts of the site through a pipeline. Distances pumped by these units have been as much as 2000 m horizontally and over 400 m vertically. Typical construction projects where such pumps are extensively used include power stations, tunnels and bridges. Schwing static pumps were used on the

Fig. 18. Typical stationary concrete pump, model BP550HDE (courtesy F. W. Schwing/ Burlington Engineers)

Fig. 19. Heavy duty trailer pump, model BRA2100HD (courtesy Putzmeister Ltd)

Fig. 20. Typical separate placing boom (courtesy F. W. Schwing/Burlington Engineers)

UK Dinorwic pumped storage scheme in the early 1980s to place 400 000 m^3 of concrete. Figs 18 and 19 show a typical stationary pump and a heavy duty trailer pump, respectively.

In certain instances, for example in power station construction, it is common practice to use separate placing booms in conjunction with large stationary pumps. A pump supplies the concrete to a boom which is positioned strategically at a central position where it can readily and economically distribute the material as required. These booms are generally the same as the types mounted on the mobile pumps with the full luffing and slewing facilities (Fig. 20).

Truckmixer concrete pumps

Truckmixer concrete pumps are combinations of truckmixers and concrete pumps, complete with short three or four arm booms. Their main application is for smaller pours, perhaps in one-off situations, and are used for reasons of access or when another method of placing is impractical. Truckmixer concrete pumps can operate more or less in the same way as a mobile pump, except that they carry the

Fig. 21. Pompuno truckmixer pump (courtesy CIFA SpA)

concrete which is to be pumped. Three and four axle chassis are used and concrete loads from approximately 3 m^3 to just under 6 m^6 are possible, depending on statutory transport regulations in the particular country. Placing outputs are up to 50 m^3/h and vertical distances of about 25 m are possible.

If additional concrete is required, a back-up truckmixer can discharge into the pump hopper and pumping can then resume. On the other hand excess concrete can be pumped back into the drum for possible use on the next site.

Although these pumps are not in the core of the manufacturers' production, and do not form a large part of pump hirers' fleets, they can provide an efficient and speedy service for specific applications where a small amount of concrete has to be placed quickly and precisely (Fig. 21).

Fig. 22. Tunnel pump with belt conveyor (courtesy F. W. Schwing/Burlington Engineers)

Fig. 23. Stationary concrete pump, model 1403HD, used at Gotthard Tunnel (courtesy Putzmeister Ltd)

Tunnel concrete pumps

Manufacturers invariably use their own range of pumps and, together with conveyors, mixing equipment and placing booms, produce various packages of what are described as tunnel concrete pumps. They have been well proven over many years. Pumps with belt conveyors and mixing equipment can give outputs in the region of 60 m^3/h (Fig. 22). Tunnel pumps are usually constructed on a rail mounted chassis but they can also be skid mounted. Other types of tunnel pumps are mobile, trailer or truck mounted types used in conjunction with two-piece placing booms. Outputs are again up to about 60 m^3/h.

Many standard pumps can be adapted for use in tunnels and in 1978 a Putzmeister high pressure static model BA1403HD pumped a then world record of 340 m vertically (a 600 m long inclined line) in the Gotthard tunnel. The pump had an increased capacity hopper (5 m^3), loaded directly from a pan mixer. The pipeline diameter was 125 mm. The delivery pressure was over 110 bar* at the greater heights, but it is claimed that the pump had sufficient reserve power to have reached 380 m (Fig. 23).

Concrete pumps for sprayed concrete or shotcrete (chapter 11) are used extensively in tunnels. These can be double piston type pumps for injection work up to pressures of 70 bar and using up to 10 mm aggregate, or pneumatic conveyors which will mix and convey concrete with up to 30 mm aggregate.

*1 bar = 101 325 Pa.

Other types of concrete pumps

Worm pumps are used principally for conveying mortar or plaster for spraying and injecting. A mixer attachment can be added if required. Delivery heights are claimed to reach ten storeys or more with favourable material. The pump is driven by a diesel or electric motor and the pumping action is surge free. When a spray gun is used, there is an automatic stop-go facility.

Road tunnels and embankments, etc., can be stabilized successfully by injecting cement slurries, mortar or sand/ cements in to them, using a worm pump. Worm pumps are simple and easy to handle, operate and clean (Fig. 24).

Mortar or plaster pumps

There are several types of mortar or plaster pumps on the market; there are single and double piston pumps as well as worm pumps. The larger double piston pumps will even pump difficult mortars up to 30 storeys high. The limit for the single piston pump is usually ten storeys. Material can be delivered ready mixed or mixed on site with an optional mixer attachment. Most have a small air compressor mounted on the pump.

Fig. 24. P11 worm pump (courtesy Putzmeister Ltd)

Fig. 25. Guni 48 Turbosol mortar pump (courtesy Sidewinder Concrete Pumps Ltd)

The material is sprayed by pumping it through a lightweight hose to a spray gun where compressed air is introduced. Most specifications can be complied with because there are different types of guns for different applications, which can be operated at varying pressures. Care must be taken in choosing the correct pump for the type of mortar or plaster being used. The manufacturers will be only too pleased to advise on this. Sprayed plastering is very much quicker than traditional plastering. A typical twin cylinder piston pump with outputs of 2–7 m^3/h is the Turbosol Guni 48, which is used for the wet shotcrete system as well as for the conveying and spraying of mortars and plasters. This unit has a maximum line pressure of 60 bar and can pump to a height of 100 m and a length of 200 m (Fig. 25).

2.2. Equipment for concrete pumping

Pipelines and couplings

The importance of pipelines and the associated couplings used in concrete pumping cannot be stressed enough. It is of little use having a pump in first class condition and able to operate well within the pumping pressures required of it, if the pipeline through which the concrete is to be pumped is inadequate to deal with these pressures. The pipeline and couplings must be of sound construction and be laid out properly (see chapter 9).

An inadequate pipeline is one which is in poor condition through wear or loose joints or which is not within the safety limits in respect of wall thickness for a given operating pressure, as advised by the manufacturers. A poor pipeline can easily cause blockages arising from leakage of grout. In concrete pumping, the very nature of the work of pushing abrasive material at high pressure through a pipeline inevitably creates a great deal of wear. The pulsations in the pipeline caused by pumping under pressure and the continual handling of the pipeline, frequent securing and releasing of couplings, creates wear at joints. All construction equipment is subject to a great deal of wear and tear. Concrete pumping pipelines, however, perhaps suffer the most, yet it is imperative that they are maintained in good order not only for making pumping easier but for reasons of safety.

Types of coupling

Couplings should be a compatible size with the mating pipe terminations. In recent years pipe terminations have been the one-piece shoulder type, with a standard nominal shoulder width of 17·5 mm and at least 6 mm high. This type should always be used and is strongly recommended by the BCPA (Figs 26 and 27).

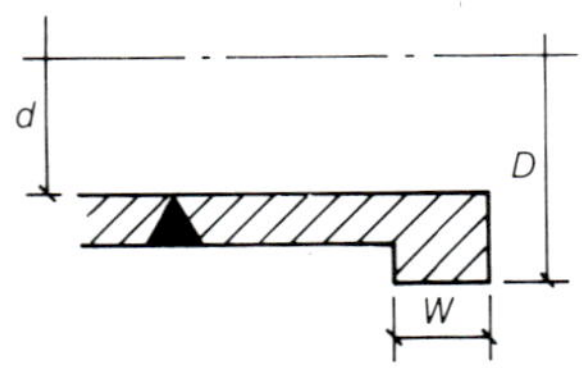

Fig. 26. Pipe coupling shoulder (courtesy BCPA)

Couplings must be of sound construction, with the seal and adaptor providing a leakproof joint, yet allowing for a small angular movement. Couplings can be either bolted or a quick release lever (toggle) type, with a facility for ensuring that the clamp is closed. This is done usually by means of a safety pin which should be fixed to a retaining wire to prevent its loss. The lever types are mostly used today. Table 3 is taken from the *Advisory code for safety in concrete pumping*[6] and is the BCPA recommendations for pipes and couplings. The BCPA recognition of safety requirements is further referred to in chapter 12.

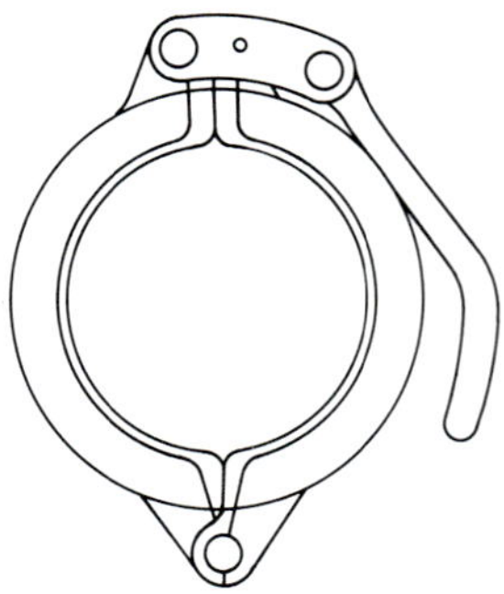

Fig. 27. Quick release type coupling (courtesy BCPA)

Pipelines

The safe working pressures and recommended test pressures of pipelines shown in Table 3 indicate the importance the concrete pumping industry attaches to this equipment. Normally the pump manufacturers will specify the pipe material, diameter and wall thickness appropriate for use with their pumps. Fig. 28 shows the minimum wall thicknesses required for various pressures and is a guide only. Advice should always be sought from the manufacturers. However, a rough guide for wall thickness requirements can be calculated from the expression

Table 3. Pipe sizes and safe working pressures (courtesy BCPA Ltd)

Normal pipeline bore *d*: mm	Nominal pipe terminating shoulder diameter *D*: mm	Nominal pipe terminating shoulder width *W*: mm	Coupling type cast/forged	Safe working pressure: bar	Recommended test pressure: bar
50	76	17·5	Cast steel	90	180
65/75	102	17·5	Cast steel	90	180
100	127	17·5	Forged steel	130	260
125	148	17·5	Forged steel	130	260

$$t = P \times d \times 3/10\,000 \qquad (1)$$

where t is the minimum wall thickness in mm, P is the pressure in the pipe in bar, and d is the diameter of the pipe in mm. Assuming that the pressure is 100 bar and the diameter of the pipe is 125 mm, $t = 3{\cdot}75$ mm, which approximately agrees with an interpolation of Fig. 28.

Wear on pipes is a constant source of worry (and expense) for the pump owner and in this respect it is worth

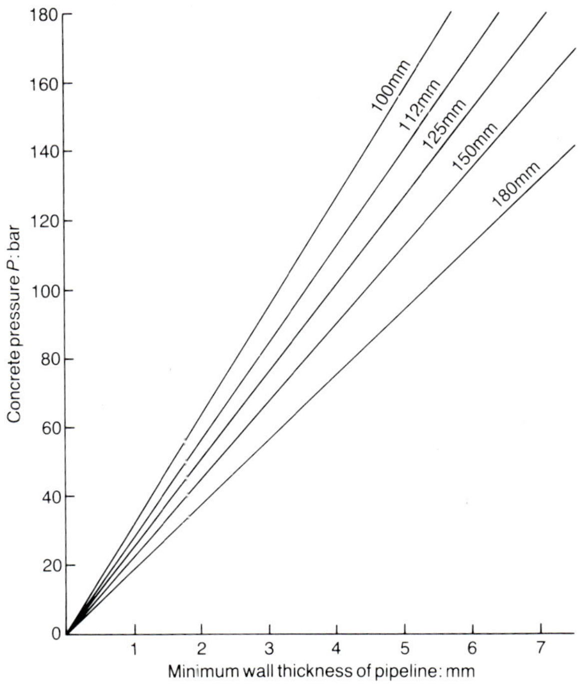

Fig. 28. Pipe sizes and corresponding wall thicknesses (courtesy BCPA)

noting that poor mixes and continual pumping at high pressures creates excessive wear. It is very important to monitor wear on pipes and this can be done by hole drilling or ultrasonic testing. Wear occurs more at bends or reducing pipes and at the bottom (invert) of the pipes. Reducing pipes or tapers that are part of the pump unit should have couplings that are capable of withstanding the maximum concrete pumping pressure.

Blockages often occur at the reducing pipe immediately behind the pump and it is essential that such blockages are cleared as quickly as possible. Therefore quick release couplings are essential.

Flexible hoses form part of the pipeline and although they are usually used at the end of the line as placement hoses, they are occasionally used to connect rigid pipes together. This practice is not to be recommended, however, as a weak link has been set up in the line and the higher frictional resistance created at this connection could lead to a blockage there. The wall frictional resistance of a flexible hose is approximately three times higher than for a steel pipe and this must always be borne in mind. Apart from the blockage nuisance there is a real danger of a serious accident if flexible hoses are used at junctions in the pipeline.

Hoses should not be subjected to pressures higher than those recommended by the manufactures and kinking should be avoided because this could cause a blockage and again raises the possibility of an accident. Lengths of end hoses should be limited to those recommended by the manufacturers. Maximum lengths of 4 m for 125 mm dia. pipes and 5 m for 100 mm dia. pipes are normally stipulated. When a hose is filled with concrete the hose is, of course, heavy and commonsense should dictate that long hoses should not be used. A 4 m length of 125 mm dia. hose filled with concrete will weigh about 110 kg which is quite enough for one person handling the hose end. For end hoses over 125 mm diameter it is almost impossible to manipulate the hose for proper placing.

Reference has been made several times here to the importance of adhering to manufacturers recommendations. This makes sense, as it should be remembered that the manufacturers' continuing improvements and refinements have in many cases arisen out of problems, and indeed accidents, in the past. It is foolhardy to ignore the information they provide, or will provide if requested.

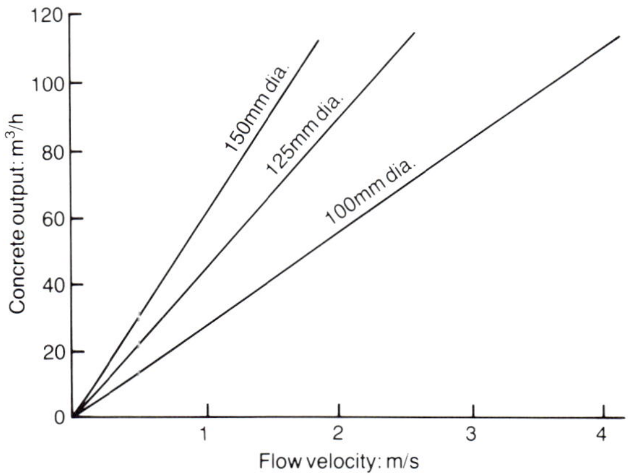

Fig. 29. Flow velocity and pipeline diameter (courtesy F. W. Schwing/Burlington Engineers)

Pipe dimensions. It should be noted that for a given pipeline diameter the flow velocity increases proportionately to the output and the pumping pressure is dependent on the flow velocity. This is illustrated in Fig. 29 and shows flow velocity as a function of pipeline diameter.

Correct diameter and wall thickness. It is important to choose the correct diameter and wall thickness of the pipeline to match the pump and the required placing rate. Nowadays almost all pumped concrete is conveyed through a 125 mm pipeline. There are exceptions, of course. For long horizontal lines involving high pumping pressures, a larger diameter pipe would be more suitable as obviously there will be less resistance to flow. For pumping concrete to heights, the fact that gravity and the weight of concrete in the line must be considered means that the smallest possible size pipe should be used.

The weight of the pipe material itself should also be considered, as the larger the diameter the heavier it becomes and the weight increases with length. The lengths of individual pipes should be limited to that which can be safely handled without causing damage or accidents.

Figure 30 shows three sizes of pipeline — 100, 125, 150 mm diameter and 3 m long, wall thickness 4 mm, with the equivalent weights when empty and full of concrete. It can be seen that when full the 100 mm dia. pipe weighs about 70 kg, the 125 mm dia. pipe about 125 kg and the 150 mm dia. pipe about 175 kg. In the event of a blockage, handling any of these pipes when filled with concrete is

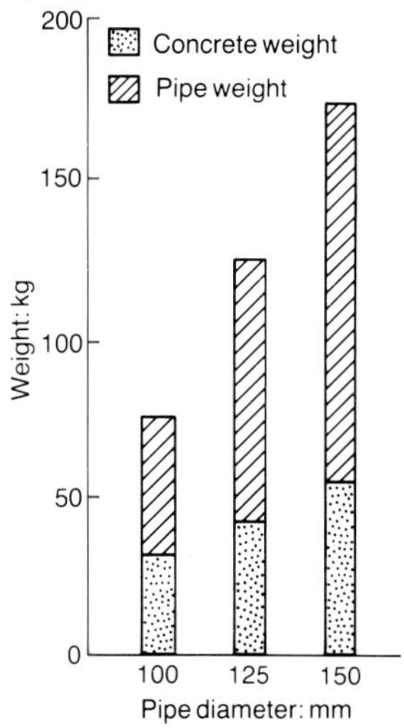

Fig. 30. Weights of pipes, empty and filled (courtesy F.W. Schwing/Burlington Engineers)

difficult, particularly the 150 mm dia. pipe. Sections of pipe are generally kept to a maximum of 3 m for this reason.

Apart from weight considerations, the use of small diameter pipelines offers advantages in easier handling of the correspondingly smaller diameter end hoses. There will also be less segregation problems if bleeding occurs and less resistance from air pockets.

Bends, taper pieces, shut-off valves, changeover valves, etc., which form part of a pipeline must be at least of similar construction to the pipeline and have pressure resistant qualities to match the pipeline, otherwise there is a weak link. It should be borne in mind that pressure is greatest at bends and taper pieces and wear here should be monitored carefully. Steps to equalize wear can include reversing the position of bends and placing worn pipes at the end of the pipeline. There is more wear on pipes nearer the pump because the pressure is higher the nearer it is to the pump. It should also be remembered that the faster the pumping is the more the wear is likely to be on the pipeline — and on the pump.

Stowing bends (S bends) and shut-off valves are sometimes used in parts of pipelines which are vertical or sloping. This is discussed more fully in chapter 9.

Other equipment

Placing booms and arms. Many of the concrete pump manufacturers have designed or arranged their range of mobile pumps and associated placing booms so that the latter can be readily used as separate placing booms, or masts as they are sometimes referred to. Further expensive development is therefore not required apart from modifications and some fixtures when installing the boom on site.

These booms are particularly suited for large structures, perhaps with a central core, where a boom can be placed which will readily cover a good proportion at least of the various areas to be concreted. Sometimes two or three booms are used, depending on the size and nature of the project. Frequently there is only one boom for several pumping positions and the boom is transferred from one position to another as required.

Separate placing booms have been used since 1969 and are without doubt an aid to faster placement and easier distribution. The criteria required to make the use of this equipment a viable proposition for the contractor includes

Fig. 31. Separate placing booms at Heysham power station (courtesy F. W. Schwing/Burlington Engineers and Taylor Woodrow Construction Ltd)

Fig. 32. Four separate placing booms at offshore oil platform construction site, Ardyne Point, Argyll (courtesy F. W. Schwing/Burlington Engineers and Sir Robert McAlpine & Sons Ltd)

such obvious items as size of contract (or volume of concrete to be poured) and layout of the structure. It is obviously not economical to erect a separate placing boom on a site for only a small amount of concrete. Where there is a considerable volume of concrete to be placed, coupled with the need to distribute the concrete in various areas, the placing boom can come into its own. For example, at Heysham power station in Lancashire, the layout of the work made it a worthwhile proposition for the contractor to use six separate placing booms at various locations (Fig. 31). At Ardyne Point, Argyll, the contractor used four placing booms to place concrete for the construction of an offshore oil platform (Fig. 32). High pressure stationary concrete pumps were used, with outputs of up to 70 m^3/h.

Separate placing booms are normally placed on tubular steel columns or structural steel columns placed around the site at carefully planned positions. Dismantling and re-erecting times are relatively short and some manufacturers claim that this can be completed in as little as 30 minutes.

Detaching a boom from a mobile pump and moving it to an independent support column is a simple operation involving the withdrawal and refixing of four bolts. Transferring the boom from one position on site to another is also simple using a crane to lift the boom (Fig. 33).

For larger contracts therefore with a high concrete content, the independent or separate placing boom can greatly facilitate the distribution of concrete. Careful planning, particularly the positioning and number of masts, is clearly important. Each contract has to be considered on its

Fig. 33. Positioning separate placing boom using a tower crane

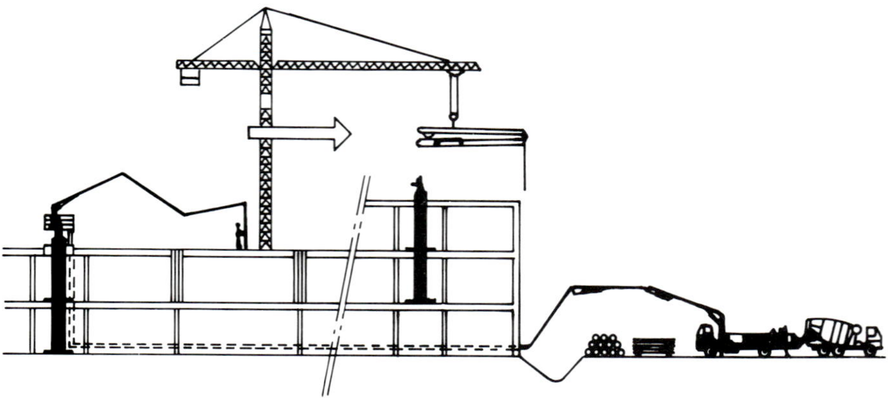

Fig. 34. Circular distribution arm (courtesy F. W. Schwing/ Burlington Engineers)

own merits and there can never be a general rule as to what will be required for a given volume of concrete or site layout.

Supporting columns can be of several types, but should always be securely anchored. Latticed framework as used for tower cranes or structural steelwork is suitable. There is always scope for ingenuity when using independent placing booms. Ingenuity can not only provide a good deal of satisfaction but also increased productivity.

Distributing arms. On a much smaller scale, various types of simple distribution arms are available as aids to concrete placing and can be extremely cost effective by saving labour (Fig. 34).

Service provided by the manufacturers

The after-sales service and spares service claimed by the manufacturers of almost any commodity will come in for criticism from time to time. Nevertheless the manufacturers of concrete pumps and equipment endeavour to live up to their claims and indeed most of them do provide a good service. The large manufacturers have branches or at least spares depots or agents in most parts of the world and their computerized systems of stock control greatly facilitate the dispatch of spare parts. Typical after-sales service will include visits to site by service engineers who routinely check performance and who will stay as long as necessary to see that the pump is working properly. In the construction industry, particularly where plant is concerned, many will agree that the expression 'going from one crisis to the next' is perhaps not too far from the truth. A breakdown does create a crisis of some sort and spares or repairs are invariably required immediately. Waiting for spares for an item of dominant plant is frustrating for site management, who will say that they have enough problems carrying out the actual construction work without having other problems such as getting an item of plant repaired quickly. The various parts of concrete pumps and equipment which are more liable to wear and damage are recognized and most pump companies are sensible enough to have spares readily available. The manufacturers also are aware of such items. With good foresight and communication, the concrete pump industry need not suffer from lack of service or spare parts.

3

Capabilities of concrete pumps

The performance and capabilities of concrete pumps continue to improve. This is not only due to development by the manufacturers, but also to advances in concrete technology. Case studies indicate the versatility of the concrete pump.

3.1. Performance outputs

The pump manufacturers claim, perhaps with justifiable authority as their performance figures are based on actual performances, that they are able to produce a concrete pump that will place concrete to the top of the highest building. The manufacturers also claim that the pump will be able to pump an excessively long distance. This may be disputed by some pump users, but the fact remains that the challenge of providing a pump to cope with a special pour of extreme height or length is one which is normally taken up eagerly by the manufacturer. Such achievements, however, cannot be readily accomplished without the co-operation and expertise of the concrete technologist.

The performance outputs, which are in m^3/h, quoted by manufacturers are the maximum theoretical outputs and they should not be considered as practical outputs. There are always other factors involved in placing concrete by pump which make it impossible for a pump to operate continuously for any significant length of time. This is explained more fully in chapter 7. The outputs given, however, are a good indication of the size of the pump which will be required for a specific pour. If the placing requirement is $30\,m^3/h$, a pump rated in excess of this, probably up to $45\,m^3/h$ or even $50\,m^3/h$, may be required. Factors such as flowability of the concrete and length of pipeline have to be considered. The highest output will be obtained from high flow, low resistance mixes pumped through a short line with few bends (see chapter 7).

Capabilities of pumps

A medium sized concrete pump with a boom or short pipeline is capable of accepting a full 6 m^3 load of pumpable concrete from a truckmixer and pumping it in five or six minutes (see chapter 1). Thus, using the higher figure, if such a supply were continuous the placing rate would be 60 m^3/h. But to maintain such a concrete supply there would have to be two truckmixers backed up to the pump hopper so that immediately the first was empty the second started discharging. Furthermore, this situation would have to be maintained for the whole of the pour if the full potential or capability of the pump were to be realized. In practice this seldom, if ever, occurs as it is not practical. It does provide, however, an example of the output capabilities of an average concrete pump under ideal conditions.

For special jobs which entail pumping over extreme heights or lengths, the manufacturers should be consulted. They are often able to recommend engine modifications such as fitting smaller diameter concrete delivery cylinders or feeding oil to the cylinder heads, which will increase the concrete pressure to give better performance. Such modifications, however, will only be fully effective with a flowing mix which is capable of withstanding high pressures without segregation or bleeding. In some of the case studies described in section 3.2 modifications were carried out.

Observations on a large pour of 500 m^3 proved that when achieving an average output of almost 50 m^3/h (which is an excellent performance), the mobile pump delivering concrete to the boom was constantly in operation, working well within its capabilities with no delays. The supply of high flow concrete was first class and the placing crew could not be faulted in their work. The pump was rated at over 100 m^3/h, yet there was clearly no evidence to suggest that the pump was in any way too powerful for the job. Most observers would have said that it was eminently suitable. Whatever the size of concrete pump in the medium to large range, the average output on what can be described as a normal pour — one which is neither difficult nor easy, neither large nor relatively small — will be around 25 m^3/h.

Concrete has been pumped to heights in excess of 400 m, but this has required the use of selected high pressure pumps and equipment together with special attention to concrete mix design. Maximum distances pumped will also depend on the pump size, diameter and condition (e.g. tightness) of the pipeline and of course the concrete mix. Lengths of over 2000 m have been pumped successfully.

3.2. Achievements and case studies

Some examples of past performances in high-rise concrete pumping, high-volume and long distance work will now be given to illustrate the capabilities of the concrete pump.

High-rise concrete pumping

In February 1985 the record for vertical concrete pumping of 432 m was achieved at the Estangento Sallente power station in the Spanish Pyrenees (Fig. 35). A Putzmeister stationary high pressure pump (BSA 2105) with an S-transfer tube valve was used. This pump has a theoretical output of 120 m^3/h, 180 mm dia. delivery cylinder and an effective concrete pressure of over 200 bar. 630 m of 125 mm dia. high pressure pipeline was required, and a gate valve was positioned close to the pump.

The concrete mix design incorporated crushed granite aggregates (which were available locally) and river sand. 0·7% Sika E90 Plastiment (relative to the cement) was added to the mix, as a high slump of 140 mm was required. Without the plasticizer the slump was 80 mm. The high slump was

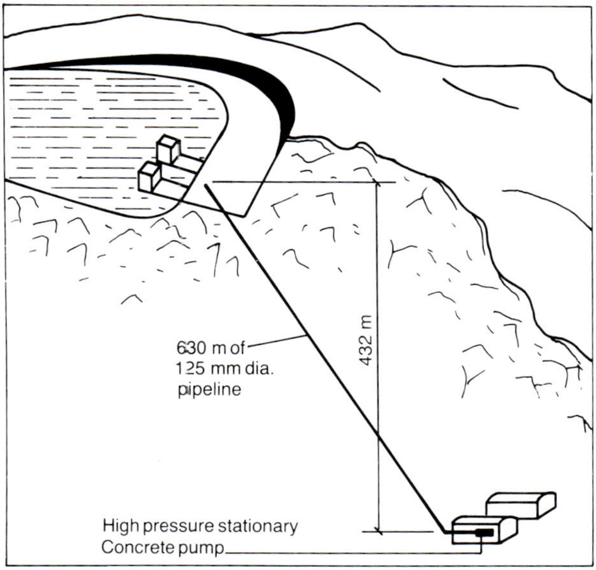

Fig. 35. World record high-rise concrete pumping at Estangento Sallente power station in the Pyrenees

necessary to fill all cavities and achieve compaction without vibrating the concrete which formed the lining to the pressure shaft. The mix consisted of 506 kg 12–25 mm granite aggregate, 362 kg 5–12 mm granite aggregate, 655 kg 0–5 mm granite sand, 360 kg 0–3 mm river sand, 211 kg cement, 90 kg fly ash and 183 l water.

A maximum concrete pressure of 170 bar was recorded with regular outputs of 20–30 m^3/h. At the final height of 432 m a pressure of 150 bar was recorded in the delivery line 1 m from the pump. The output then was 21 m^3/h.

The new Post Office Tower in Frankfurt-am-Main broke the world record for the highest concrete pumping in 1977, when concrete was pumped 310 m vertically (Fig. 36). Previously, high speed cranes usually equiped with high capacity winches had been used for such work. Concrete requirements were 80–120 m^3 in a 2·44 m rise per day. The actual placing time was restricted to a three hour period and so relatively high outputs were required. A Putzmeister Elephant model BRS 2104E with a 132 kW electric motor was used (Fig. 37). This pump has a long piston stroke of 2100 mm, and a dampening hydraulic control.

A smooth pumping action is very important in concrete pumping at great heights. For example, at 200 m the vertical column of concrete weighs up to 13 t and, at the end of the piston stroke, it creates a tremendous feedback force on to the piston head. Putzmeister claim that their dampening control largely cancels this out. A major problem when pumping at heights is to restart the upward movement of concrete at the beginning of each stroke.

Fig. 36. Frankfurt Post Office Tower (courtesy Putzmeister Ltd)

Fig. 37. Putzmeister 2100 series concrete pump (courtesy Putzmeister Ltd)

Thick walled 125 mm dia. high pressure pipes were used and tested to 300 bar. Each length of steel pipe was securely fastened to the concrete wall. Two shut-off gate valves were installed at the bottom of the vertical section, one for emergency shut-off and cleaning and the other for draining the line at the finish of the day's pumping. There was a horizontal start length of approximately 30 m from the pump to the bottom of the vertical section.

To facilitate distribution at the placing position a separate placing mast was used. Pumping pressures ranged from 22 bar at 100 m to 42 bar at 200 m and at 310 m the pressure exceeded 100 bar (Fig. 38).

The concrete mix, which included a plasticizer, was of

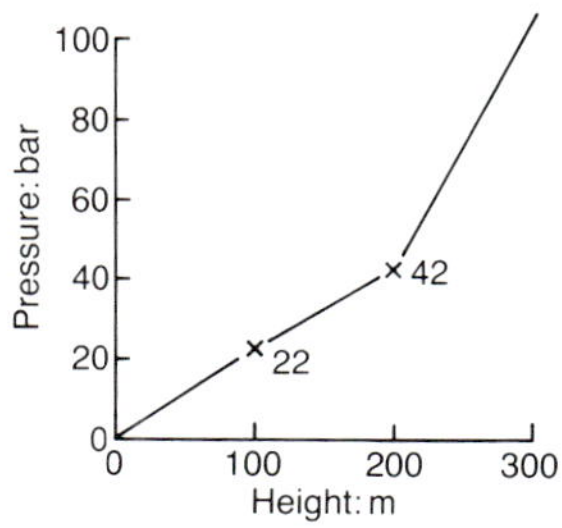

Fig. 38. Pumping pressures at Frankfurt Post Office Tower (courtesy Putzmeister Ltd)

Fig. 39. Separate placing boom at Dresdner Bank, Frankfurt (courtesy F. W. Schwing/ Burlington Engineers)

natural aggregates of maximum size 32 mm and the cement content was 420 kg/m^3. The slump was approximately 100 mm.

The Dresdner Bank in Frankfurt is one of the tallest buildings in Europe, consisting of 32 floors (Fig. 39). It was built in 1976 and concrete was pumped to a height of 170 m, which caused a good deal of interest at the time. Today this would be considered to be commonplace, as heights in excess of 100 m are regularly reached.

The pump used in this instance was a Schwing single rod gate valve stationary pump BP 550HDE with a theoretical output of 47 m^3/h and maximum concrete pressure of 101 bar. A separate placing boom with a horizontal reach of 19 m was installed for distributing the concrete (Fig. 40). Table 4 shows the height, quantity of concrete, average output and concrete pressure. These figures are indicative of typical performances by a concrete pump under such conditions.

*Table 4. Quantities and outputs of concrete at the Dresdner Bank, Frankfurt (courtesy F. W. Schwing GmbH)**

Pumping height including placing boom: m	Construction section	Quantity of pumped concrete: m^3	Average output m^3/h	Concrete pressure: bar
105	Floor slab I 16th floor	200	30	55
105	Floor slab II 16th floor	244·5	30	55
115	North core 18th floor	209	28	57–63
115	Floor slab II 18th floor	239	28	50–52
125	Floor slab I 20th floor	200	28	50–54
125	Floor slab II 20th floor	259	28	50–54
135	Floor slab I 22nd floor	200	28	58–65
135	Floor slab II 22nd floor	259	28	52–58
145	Floor slab I 25th floor	201	28	64
145	Floor slab II 25th floor	251	28	61
157	Floor slab I 29th floor	205	24	72
157	Floor slab II 29th floor	264·5	24	64–72
160	Floor slab I 30th floor	187	24	64–72
160	Floor slab II 30th floor	260	24	64–72
164	Floor slab I 31st floor	248	23	72
164	Floor slab II 31st floor	322	23	72
170	North core 32nd floor	197	21	72–75
170	South core 32nd floor	209	21	72–78

* Concrete mainly used was as follows. Cement: 380 kg/m^3 PZ 450F; admixture: 0·4% Sika BV Sp.; water/cement ratio = 0·5; consistency: K_3; aggregates: 0–2 mm, 40%; 2–8 mm, 20%; 8–16 mm, 40%.

Fig. 40. Separate placing boom (KVM22) at top of Dresdner Bank, Frankfurt (courtesy F. W. Schwing/Burlington Engineers)

High volume work

In the past decade the Dinorwic underground storage electricity power station was constructed in North Wales for the Central Electricity Generating Board. The power station is the largest of its kind in Europe. More than 400 000 m^3 of concrete was placed by pump over several years, principally as tunnel linings, varying in thickness from 300 mm to 1000 mm.

The size of the tunnels enabled lorry mounted pumps to be used, which gave increased mobility in the tunnels. These mobile concrete pumps were supplied almost exclusively by Schwing together with portable pumps and separate placing booms (Fig. 41). The large quantity of concrete pumping equipment was required to meet the high daily demands of concrete production on this project. Pumping distances close to 500 m were normal using Schwing BP550 series pumps with 125 mm dia. pipelines. The sheer size alone of the Dinorwic project makes it an outstanding example of civil engineering construction under difficult conditions.

After the disaster at the Chernobyl nuclear power station in 1986, concrete pumps played an important role in the

Fig. 41. Mobile tunnel placing boom used at Dinorwic (courtesy F. W. Schwing/Burlington Engineers)

remedial work of encasing the damaged reactor in 400 000 m^3 of concrete. More than ten concrete pumps from four different manufacturers were used. Additional operators had to be trained to ensure day and night working. Only two hours exposure to radiation was permitted per worker; each worker had a radiation counter which provided an early warning if radiation levels were too high. A camp 120 km away was set up to ensure that there were sufficient rest periods for the labour force. Concrete pumping continued day and night.

Speed was essential in the encasement of the reactor and it is interesting to note that the main concrete placing option chosen was by concrete pump. This is an example of the reliability, durability and high output performance of the concrete pump, which would not have been used in the emergency at Chernobyl without such qualities.

Putzmeister supplied most of the concrete pumps used at Chernobyl, including its giant mobile M52 with its 52 m

Fig. 42. Putzmeister mobile pump with 52 m boom ready to leave for Chernobyl (courtesy Putzmeister Ltd)

Fig. 43. Pumping concrete at Chernobyl (courtesy Putzmeister Ltd)

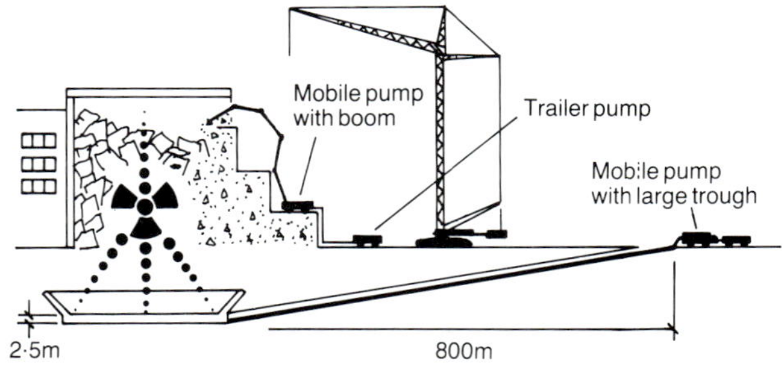

Fig. 44. Pumping concrete through a tunnel at Chernobyl (courtesy Putzmeister Ltd)

boom. The units were specially equipped to protect the operator from radiation (see chapter 2). Other precautions were also taken to ensure the safety of the work force.

Figure 42 shows one of these mobile pumps before leaving for Chernobyl. A pump similar to this one pumped 100 000 m^3 of concrete in three months. Fig. 43 shows several mobile pumps (two in tandem) placing concrete into the damaged reactor at Chernobyl. Heavy duty trailer pumps were also used. Fig. 44 shows the pumping of concrete through an 800 m long tunnel to fill a 2·5 m deep cavity under the reactor building to seal off any possible ground contamination.

Concrete pumping to extreme lengths

During the construction of Kuhtai power station in Austria, concrete was pumped through long pipelines for a 1477 m long waterway. Fig. 45 shows how pumps were used at three different locations. As can be seen, pump 1, which

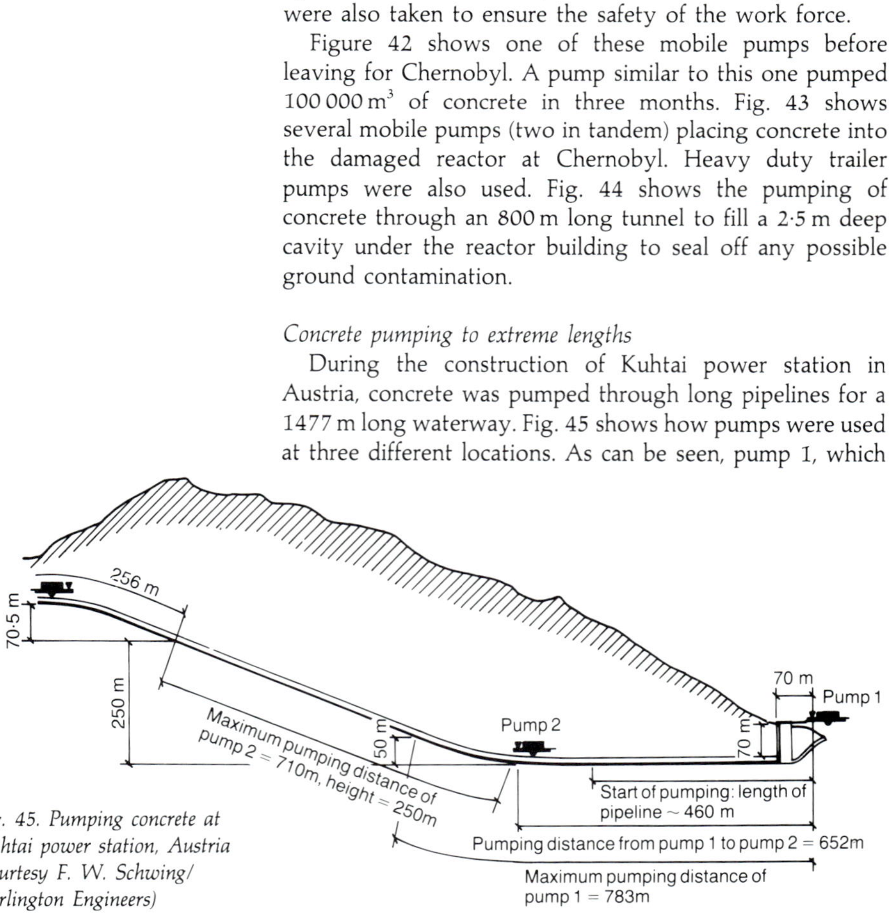

Fig. 45. Pumping concrete at Kuhtai power station, Austria (courtesy F. W. Schwing/ Burlington Engineers)

Table 5. Contents of West German standard B30 grade concrete

Crushed aggregates (grandiothric stone)	
0–1 mm	28%
1–4 mm	18%
4–12 mm	5%
12–16 mm	7%
16–32 mm	42%
Water/cement ratio	0·625
Cement	360 kg/m^3
Slump	130 mm
Liquefier Sedikament (Sika)	3·6 l/m^3

was positioned at the concrete mixing plant, had to pump 70 m horizontally and then 70 m vertically down a shaft, before pumping 652 m horizontally to pump 2. A gate valve was installed at the bottom of the shaft for cleaning purposes. Pump 1 was a standard Schwing BP550HDE stationary pump with a 110 kW motor, 150 mm dia. cylinder, output of 40 m^3/h and maximum concrete pressure of 101 bar. The second stage pump, which had to pump as high as possible up the incline, was Schwing's very powerful BP3000HDE-15 which has a 132 kW motor, capable of producing a pressure of 200 bar in the concrete, and a theoretical output of 90 m^3/h. Concrete was in fact pumped 710 m and up to a height of 250 m.

The remaining 256 m was pumped from the top by pump 3, with special precautions being taken to avoid the concrete running away. This was done by installing a shut-off valve near the pump which could be hydraulically actuated in emergencies. When pumping downhill, it was also important to keep the pressure high enough to exceed the natural flow of the concrete.

A standard 125 mm dia. pipeline with a wall thickness of 4 mm was used on the initial length from pump 1 to pump 2 but with increased concrete pressure on the inclined section the wall thickness was increased to 7 mm. Hydraulic shut-off valves which could be operated by remote control were placed in the pipeline near the delivery point into the formwork.

The concrete mix specified high early strength to enable formwork stripping after eight hours. The concrete mix was German Standard B30 grade: the details of the mix are given in Table 5.

A 12 m long section containing 70 m^3 of concrete was

Fig. 46. Schwing BP3000HD heavy duty concrete pump prepared for delivery to Sizewell B power station (courtesy F. W. Schwing/Burlington Engineers)

completed in an average time of 3·5–4 hours and the maximum concrete pressure on stage 2 was 140 bar. Considerable planning and expertise was required for the whole operation, particularly in the anchoring of the pipeline, the positioning and operation of the shut-off valves and the monitoring of hydraulic and concrete pressure (Fig. 46).

The case studies show some of the capabilities of concrete pumps. There have been many other outstanding achievements in concrete pumping in the past and there will no doubt be more in the future.

4

Pumpable concrete

To pump concrete the pressure exerted by the pump on the concrete has to overcome the pipeline wall friction, the resistance created at bends and tapers, the concrete pressure head when placing at a level higher than the pump, and the inertia of the concrete in the pipe.

4.1. Mechanics of concrete pumping

A clear understanding of what happens to concrete when it is pumped through a pipeline is fundamental to any study of concrete pumping and when designing a pumpable mix (see section 4.3). Ede[7] and Weber[8] carried out much of the early work on the behaviour of concrete in a pipeline and their published information is still relevant today. During the 1970s further investigations by Loadwick[9] and Browne and Bamforth[10] into the mechanics of concrete pumping were made and test methods for pumpability were developed.

Loadwick[9] describes the physical flow state of concrete in a pipeline as follows (Fig. 47). Pumpable concrete emerging from a pipeline flows in the form of a plug which is separated from the pipe wall by a thin lubricating layer consisting of cement paste. The water in the paste is hydraulically linked with the interparticle water layer in the plug. According to

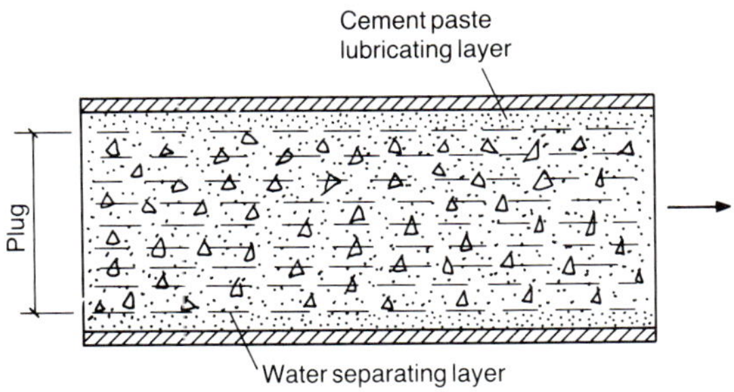

Fig. 47. Model for concrete state in a pipeline

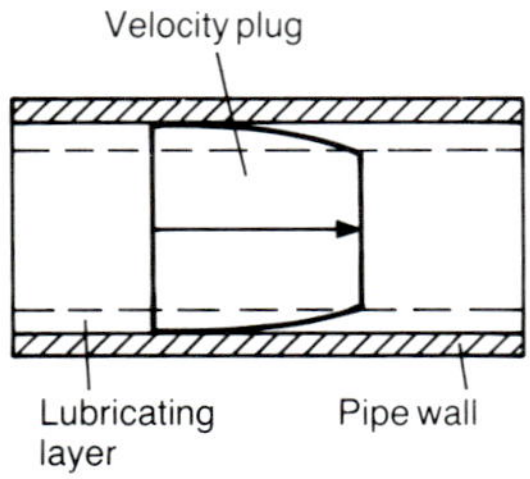

Fig. 48. Velocity profile for plug flow

hydraulic theory the plug flow velocity across the pipe is constant across the width of the plug, i.e. there is no relative velocity between the aggregate particles (Fig. 48). The velocity drops across the lubricating layer to zero at the pipe wall. The flow resistance is shown to be constant with respect to pipeline length and the concrete pressure decreases linearly down a straight horizontal pipeline. This hydraulic condition was shown by Ede[7] to be applicable to concrete only in a saturated state, i.e. a concrete containing more water than that which is required to fill the voids. The flow resistance is given by

$$R = \frac{D}{4}\frac{dP}{dx} \tag{2}$$

where R is the flow resistance (force per unit internal surface area of pipe), D is the pipe diameter and dP/dx is the pressure gradient. R consists of two components: the hydraulic shearing of the lubricating layer and the friction of solid particles on the pipe wall. Both of these components are affected by the degree of saturation of the concrete. Ede shows (Fig. 49) how the water content affects the flow resistance.

For continuous plug movement, the pressures generated by the flow resistance must not be greater than the pump

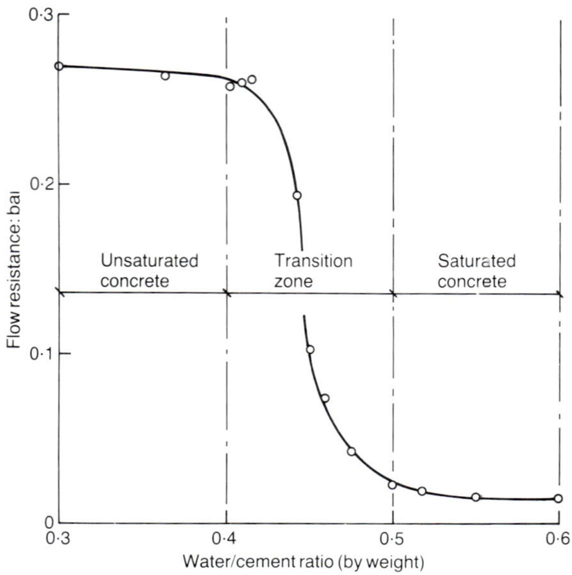

Fig. 49. Effect of water/cement ratio on initial flow resistance

Fig. 50. Effect of dewatering in a pipeline

pressure rating. However, if the concrete is too saturated the permeability of the concrete at certain pump pressures may be such that water is forced out of the mix (i.e. bleeding), creating an increase in flow resistance and a possible blockage. Fig. 50 illustrates such a condition.

Therefore to enable concrete to be pumped, both the permeability and the flow resistance, which are functions of the mix proportions, have to be carefully considered. The permeability is affected by the pressure gradient, which is directly proportional to the flow resistance.

Tests for pumpability

Browne and Bamforth[10] describe in detail three test methods for measuring pumpability, namely the pressure bleed test, measurement of voids, and measurement of concrete pressure in the pipeline.

The pressure bleed test is a static test to simulate the state of concrete under pressure in a pipeline, in which the dewatering characteristics (or permeability) of the concrete can be measured. The apparatus is illustrated in Fig. 51. A sample of concrete is compressed into a 125 mm dia. cylinder and the quantity of mix water which drains out under sustained pressure is measured. A pressure of 35 bar is applied by a hand pump and the quantity of water emitted is measured at specific time intervals while maintaining the pressure on the concrete. Typical bleed test results are shown in Fig. 52. In the results shown in Fig. 52, it was found that the volume of water emitted after the first 140 s was small and the test was discontinued after this time. Water remaining in the sample exists in the voids between compressed particles. If the initial water content is known, the residual water content can be used as a measure of the compressed void content of the solid particle system.

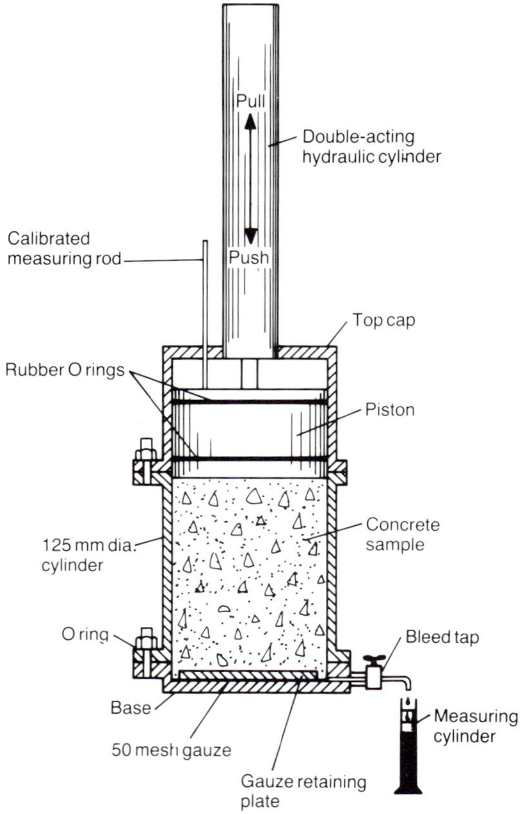

Fig. 51. Pressure bleed test apparatus

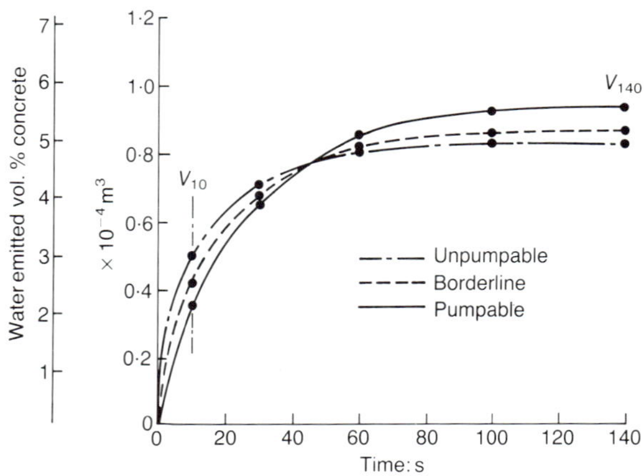

Fig. 52. Typical bleed test results

A concrete prone to dewatering will emit water quickly during the first 10 s and the volume of water emitted between 10 s and 140 s will be small. Higher values indicate that the concrete can be pumped more easily. Fig. 53 gives an interpretation of bleed test results. If the results for the time between 10 s and 140 s are less than the permissible values previously established for different levels of workability, the mix will have to be modified.

A void measuring apparatus was developed by Kempster[11] as shown in Fig. 54. In this test the head of water which can be supported by a partial vacuum created within the aggregate is measured. Layers of aggregate are hand compacted to 35 mm below the top of the sample jar and the airtight lid is screwed down. The drain tap is then opened. With the reservoir in position 1, the water level in the measuring tube is brought to a predetermined level by adjusting the water level in the reservoir. The tap is closed and the apparatus is now set up as shown in Fig. 54. The voids are measured by lowering the reservoir to position 2 which creates a pressure head in the measuring tube. The head is inversely related to the volume of air in the system.

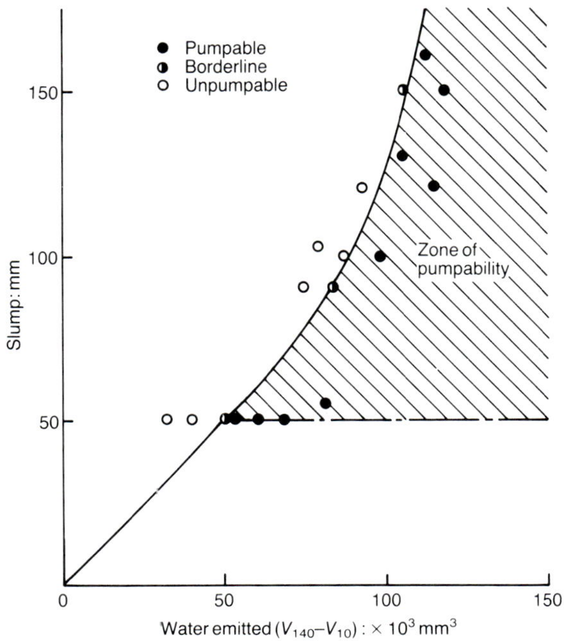

Fig. 53. Interpretation of bleed test results

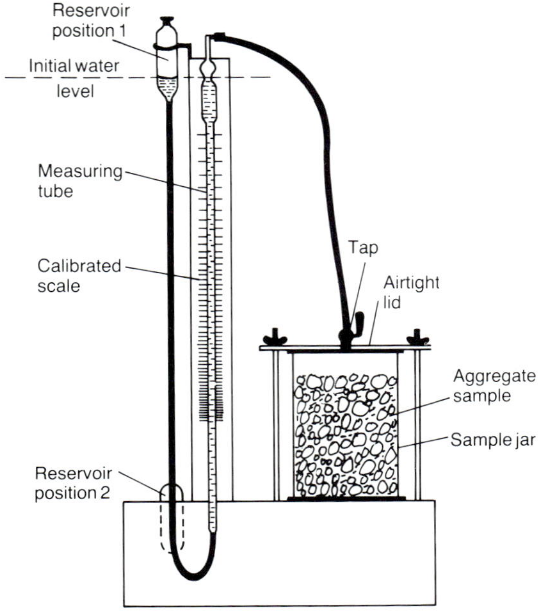

Fig. 54. Void measuring apparatus

The void content can be read directly from the calibrated scale.

Figure 55 illustrates the typical effect of varying the sand content on the voids in the combined aggregate and also the voids of the total solid particle system when the cement content is increased above an optimum level. The void meter is not only useful for assessing aggregates for pump mixes but also for assessing effectiveness of other void filling materials such as fine sand, pulverized fuel ash and ground granulated blast-furnace slag.

A method for measuring in-line pressure has been developed for use in the laboratory or on site, to record indirectly the pressure of concrete in a pipeline which can give an indication of pumpability and variability of the concrete as it is being pumped. The arrangement is shown in Fig. 56. The line pressure is measured by the strain developed in a short section of calibrated pipeline. Results are recorded in terms of voltage change from a bridge of electrical resistance strain gauges positioned on the pipeline. Results of a pressure trace from a pen recorder which was operated at 1 mm/s are shown in Fig. 57. When the recorder is operated at 10 mm/s more detailed information is available including concrete pressure, concrete velocity and efficiency of the pump. An analysis of a typical pressure pulse is shown in Fig. 58.

Browne and Bamforth[10] carried out a series of tests to measure the variability of site mixed concrete, the results of which are shown in Fig. 59. Readings on a strain gauge were taken on three occasions for periods of up to four hours and the set of traces shown indicate the variation in workability with the corresponding differences in throughput and flow resistance.

Thus the recording of in-line pressure/time traces from a

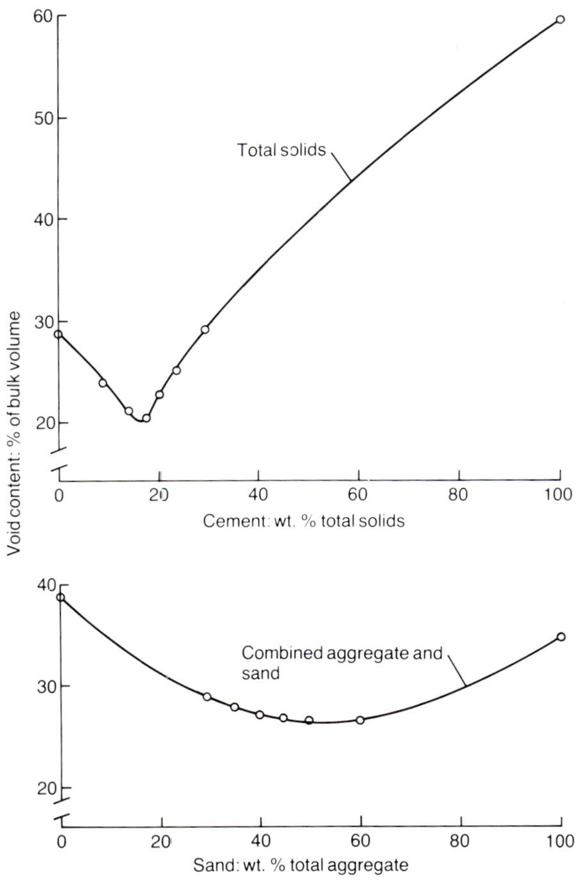

Fig. 55. Typical values of void content

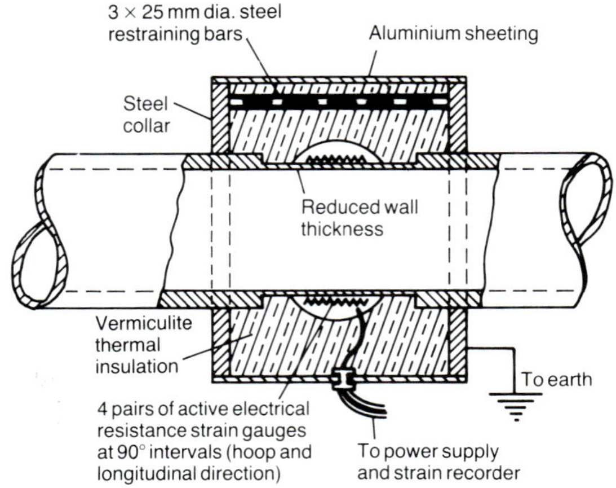

Fig. 56. Instrumented pipeline

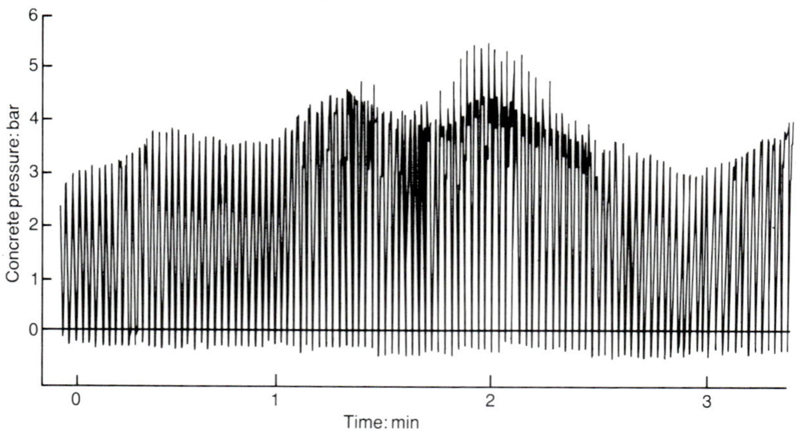

Fig. 57. Continuous pressure trace

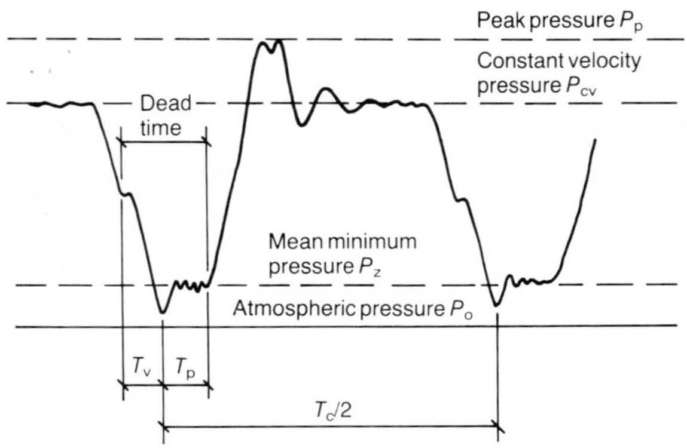

Fig. 58. Analysis of pressure pulse

strain-gauged pipeline provides a method of measuring the effectiveness of different types of pump and improvements to the pumping process as well as observing variations in materials and mix proportions.

4.2. Design considerations for a pumpable mix

With permeability and flow resistance in mind, the mix must be able not only to bind all the constituent materials together under pressure from the pump, thereby avoiding segregation and bleeding, but must also facilitate the radial movement of sufficient grout to maintain the lubricating film initially placed on the pipeline wall. The mix should also be able to deform (i.e. alter its cross-sectional area) while flowing without causing segregation. The mix deforms in the valve assembly, at bends and taper sections. To achieve this, the proportion of fines, i.e. cement and fine sand below 0·25 mm size, is of prime importance and quantities of between 350 and 400 kg/m^3 are considered necessary for pumpable concrete. Not only is this required for maintaining the lubricating film, but it is important for quality and workability that there is enough grout to cover individual grains.

Assuming that the pump is mechanically sound there are two main reasons why blockages occur. The plug of concrete will not move because either

(*a*) water is being forced out of the mix creating bleeding and blockage by jamming, or
(*b*) there is too much frictional resistance due to the nature of the ingredients of the mix.

Both conditions are caused by poor grading at opposite extremes. Condition (*a*), where high voids are involved, increases flow resistance locally, and condition (*b*), where there is a high proportion of very fine materials, increases flow resistance generally. A report by the Concrete Society[12] included a diagram which illustrates this (Fig. 60). Recommendations were made to counteract

(*a*) excessive frictional resistance by: decreasing the cement content; adjusting the aggregate coarse/fine ratio to increase voids; increasing water content; and using a wetting agent admixture
(*b*) excessive segregation and bleeding by: increasing the cement content; adjusting coarse/fine ratio to reduce

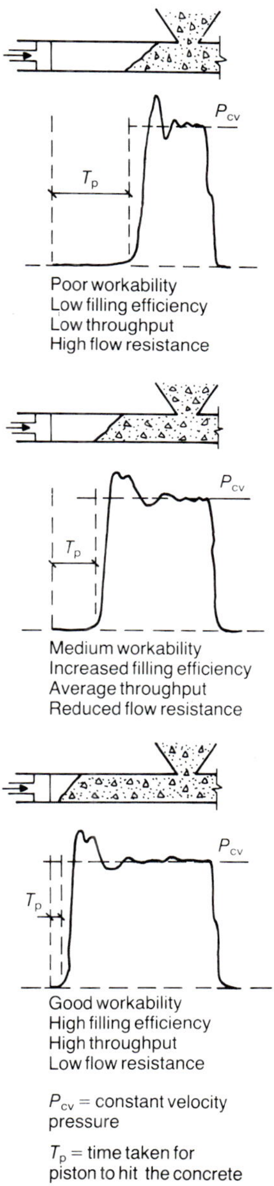

Fig. 59. Effect of concrete workability on pumpability and pump efficiency

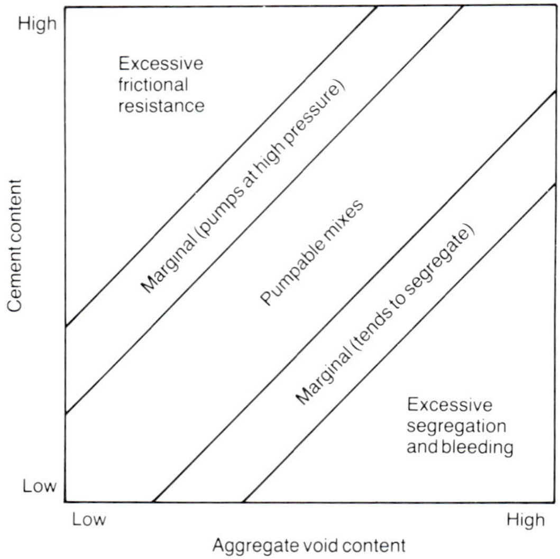

Fig. 60. Relationship between cement content, aggregate void content and excessive friction on segregation and bleeding

voids; adding fines (e.g. rock dust or pulverized fuel ash); and using a flocculating agent admixture.

Kempster[13] describes the blocked filter effect of fines which allows the water phase to transmit pressure but not to escape from the mix. Yet at the same time there is high frictional resistance created by these fines. In designing a pumpable mix therefore, the aim must be to produce a concrete suitably graded to create this blocked filter effect and to maintain the layer of cement paste at the pipe wall.

Grading and void content

To obtain the correct grading and void content the mix designer's priority must therefore be to have a proper understanding of the available aggregates. This is a prerequisite for pumpable mixes or indeed for any other mix.

Aggregates vary in shape, grading and porosity according to their source and may in fact vary from the same source. The types and size of aggregate are extremely important, particularly when designing a pumpable concrete. At the same time it should be remembered that concrete has to be designed to a specification to suit different requirements and, for economic reasons, the concrete must be made from materials which can generally be obtained locally. Such conditions render it impractical to be specific about grading figures which would encompass pumpable concrete in general. There are many sands and gravels with natural gradings that give low void contents and facilitate a pumpable mix design. Because of the different shapes of aggregates, mix proportions will differ for pumpable mixes using different aggregates. As natural gravels are usually rounded and a wide range of sizes are available, they produce better pump mixes than crushed rock aggregates. On the other hand, crushed rock can be obtained in a wider range of single sizes from which gradings can be readily produced. Coarse crushed materials may contain a proportion of dust, an excess of which could cause high pipeline friction.

It is not always possible, of course, to design a pumpable mix around an ideal aggregate for the job. Sometimes there will be no choice as types of aggregates vary in each part of the country. Naturally occurring round or flaky (elongated) aggregate as well as crushed aggregates are suitable for a pumpable mix, but it is essential to be aware of the differences in size and composition. These should be allowed for in the mix design.

Reference has been made previously to the requirement of a sufficient quantity of fines and grout, bearing in mind that each grain (both large and fine) should be covered by grout. It should be remembered that the surface area of a piece of aggregate of given volume is much less than several pieces of aggregate which make up the same volume as the single larger piece. Similarly an elongated or a crushed piece of aggregate has a greater surface area than a rounded piece of the same volume (Fig. 61).

The mix designer must be thoroughly aware of the nature of the materials to be used and from this knowledge the designer will be able to produce suitable gradings for a specific mix.

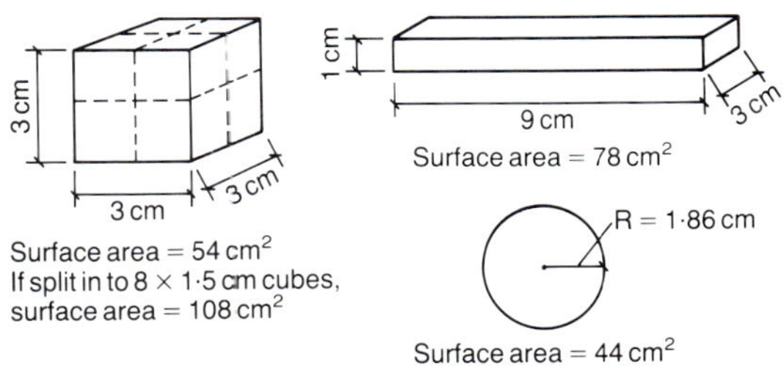

Fig. 61. Effect of aggregate shapes on mix design

4.3. Designing a pumpable mix

When ordering concrete for a pumped pour, the ready-mixed concrete supplier will invariably need to know if the concrete is to be pumped, as it is essential that the mix design takes this into account. Yet in many cases the contractor fails to inform the supplier, when ordering the concrete, that the material is to be placed by pump.

The mix designer has to bear in mind the mechanics of concrete pumping as described earlier, because they are extremely important. Not only has the concrete to meet specification requirements of strength and other properties, but in its plastic state it must be pumpable. Pumpable quality means that the consistency must be such that it will pump without segregation or bleeding, that the slump will normally be within the 75 mm–collapse range and that the diameter of the pipeline is at least 3–4 times the maximum size of aggregate. These are the main criteria therefore of the pumpable mix.

The specification will include a compressive strength requirement and design for this requirement should be carried out initially in the usual manner, bearing in mind the constraints already indicated. The next step is to check the grading of the aggregates to determine the void content. This will provide the information for adjustment of the proportions of coarse and fine aggregates to make up the even grading which is necessary for a pumpable mix. Whatever the method of placement, a sieve analysis should be carried out to ascertain whether there is an excess or deficiency in certain sizes.

The grading of the aggregates should now be checked against grading curves which are known to give pumpable mixes from the same source. Typical grading curves are illustrated in Fig. 62. The aggregates of the proposed mix should have a grading parallel to these curves, but not coarser than curve 2. Adjustments to bring the grading parallel to the curves can be made by altering the proportions.

It is recommended that 10–20% of the fine aggregate should pass the 300 μm sieve. Some designers add up to 3–4% of sand as a safeguard against undersanding. Von Eckardstein[14] recommends certain quantities of fines, i.e. cement and sand in the range 0–0·25 mm, with varying sizes of aggregates. This is reproduced in Table 6.

When a suitable grading and void content has been

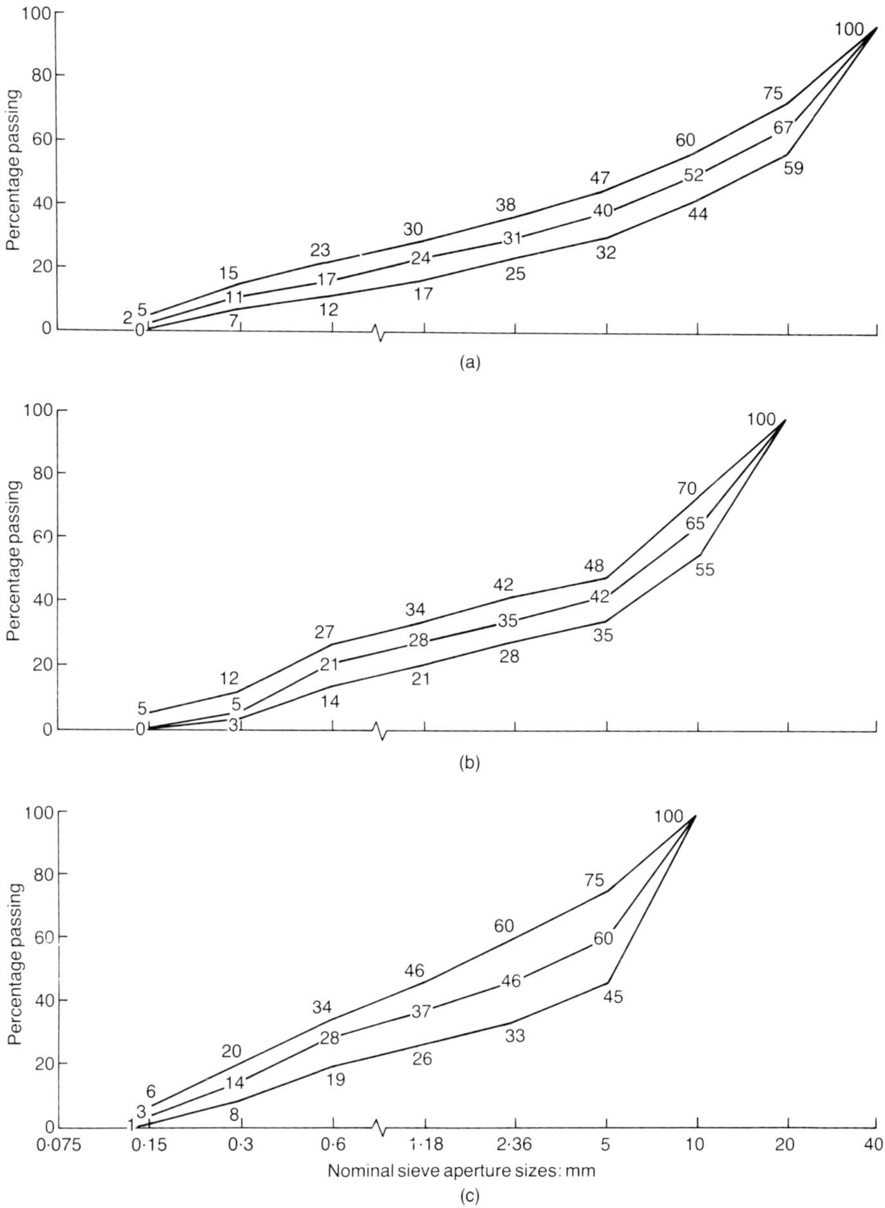

Fig. 62. Typical grading curves for pumpable mixes

Table 6. Fines quantities and aggregate sizes (courtesy F. W. Schwing GmbH)

Maximum size grain of the aggregate mix: mm	Content of fines per cubic metre of concrete: kg
8	525
16	450
32	400
63	325

established, the amount of finest fines (or cement) to fill the voids in the aggregate and to cover each grain, has to be calculated. As a rough guide, the minimum amount of cement required for a 20% void content would be

$$0{\cdot}2 \times \text{weight of } 1\,\text{m}^3 \text{ of cement (say } 1450\,\text{kg)}$$
$$= 0{\cdot}2 \times 1450 = 290\,\text{kg cement.}$$

These are the basic steps to be taken when designing a pumpable mix, but a check should now be made that the final mix proportions which have been calculated will produce the specified strength and other requirements. It may also be necessary in certain circumstances, and indeed usually desirable, to carry out trial mixes. This, of course, is sometimes specified. ACI reports[15,16] include useful information on mix design and recommendations for proportioning pumpable concrete, optimum grading of aggregates, cement, water and admixture requirements.

It is not possible to stipulate specific gradings, quantities of constituent materials or rules for pumpable concrete mix design. It should be remembered that concrete is produced from many different aggregates from different sources, varying in shape, grading and porosity. Mix design has to cater for all such vagaries in the materials available. Each case has to be treated individually.

Basic design calculations for a pumpable mix

The following example of simplified design calculations for a pumpable mix will illustrate one procedure which could be adopted.

Figure 63 shows grading curve limits for pumpable mixes using 20 mm maximum size aggregates. It is important that there is continuity of grading between sizes, particularly between 10 and 1·2 mm. The grading for the proposed mix should fall within the curve envelope as indicated in Fig. 63. If necessary the proportions of fine and coarse aggregate can

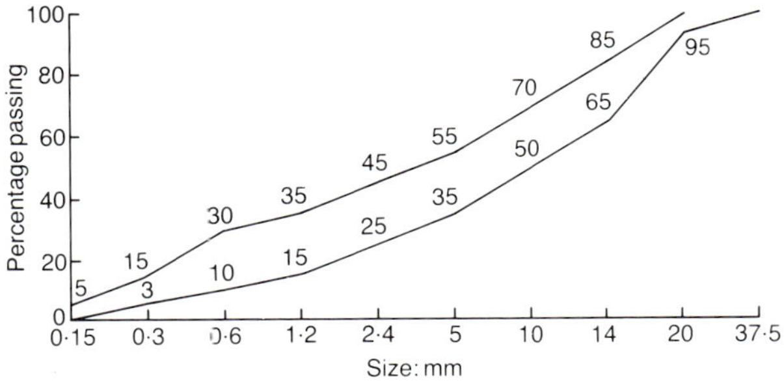

Fig. 63. Grading curve envelope for pump mix

now be adjusted. The limits of the fine solids content should be as given in Table 7.

Mix proportions on a dry basis for a 75–100 mm slump concrete are as given in Table 8. The aggregate gradings and absolute density of materials are given in Tables 9 and 10, respectively. The calculation of grading is given in Table 11, based on the proportions of 880 kg fine aggregate and 930 kg coarse aggregate. These combined grading figures fall within the limits indicated in Fig. 63.

The fine solids content should then be checked. The cement content in this particular example is 310 kg. As the fine solids content is 5·0%, the amount of fine aggregate is

$$5{\cdot}0\% \times 880 \times 3100/2650 = 51\,\text{kg}.$$

Therefore the total fine solids content is 310 + 51 kg = 361 kg. This is within the limits (Table 7) of fine solids for 190 l of water.

*Table 7. Limits of the fine solids content**

Free water content: l/m³	Fine solids: kg/m³	
	Minimum	Maximum
150	260	365
160	280	390
170	295	415
180	315	440
190	330	465
200	350	490
210	365	515
220	385	540
230	400	565
240	420	590

* Calculated on an absolute density of 3100 kg/m³.

Table 8. Mix proportions for 75–100 mm slump

Substance	Quantity: kg
Aggregate	
Fine, zone M	880
Coarse, 5–20 mm	930
OPC	310
Water	190
Total	2310

In the event of design calculations producing a marginal mix which in practice may not cater for any variations in fine material and/or water content, it would be prudent to increase the cement content slightly and also the water content. This would not affect the strength. Alternatively, the use of a plasticizer would reduce the water content. As a guide, 10 l of extra water increases the slump by approximately 30 mm and the concrete strength will be roughly 4 N/mm^2 less at 28 days.

Natural gravels, which are rounded and often have gradings that give low void contents, are very suitable for pump mixes. However, crushed rock suppliers offer a wide range of single size aggregates from which required gradings can be readily produced. These aggregates usually require higher sand contents to achieve the required void content. Crushed rock fines contain a lot of material that passes the 150 μm sieve and care must be taken to avoid an excess of this size which could cause high pipeline friction and blockages. However, properly designed mixes are possible using either natural gravels or crushed rock.

Table 9. Aggregate gradings

Sieve mesh size: mm	Percentage passing	
	Coarse	Fine
20	100	
14	60	
10	37	100
5	6	94
2·4	1	84
1·2		70
0·6		53
0·3		18
0·15		5

Table 10. Absolute density of materials

Material	Density: kg/m^3
OPC	3100
Fine aggregate	2650
Coarse aggregate	2550

4.4. Admixtures

The use of admixtures (to BS 5075)[17] can be readily divided into two areas when concrete pumping is being considered

(*a*) admixtures which facilitate the pumping of the mix, e.g. workability, water retention
(*b*) admixtures which are able to produce specific requirements, e.g. durability.

There is also the overriding factor of economics and the mix designer obviously has to bear this in mind constantly. It is of little value to use an admixture simply to make a certain property that is a design requirement of the concrete when a redesign of the ratio of normal (and cheaper) ingredients can bring this about more cheaply. But in many cases the use of admixtures can be an economic proposition and will also improve the quality of the concrete.

Reverting to the basic requirements of a pumpable mix as previously described (see section 4.2), the mix must be capable of resisting bleeding and segregation under pressure from the pump and at the same time maintain an annular film of grout at the walls of the pipeline. Admixtures which assist with this, or at least do not detract from this, can be beneficial.

Although there are many types of admixtures available, the six main types that are widely used in pumpable mixes are

(*a*) water-reducing admixtures

*Table 11. Calculation of aggregate grading**

Sieve mesh size: mm	Percentage passing		
	Coarse	Fine	Total
20	51·5	48·5	100
14	31·0	48·5	79·5
10	19·0	48·5	67·5
5	3·0	45·5	48·5
2·4	0·5	41·0	41·5
1·2	—	34·0	34·0
0·6	—	25·5	25·5
0·3	—	9·0	9·0
0·15	—	2·5	2·5

* Calculation based on 880 kg fine aggregate and 930 kg coarse aggregate.

(*b*) water-thickening admixtures
(*c*) air-entraining admixtures
(*d*) accelerators and retarders
(*e*) blends of admixtures and pumping aids
(*f*) other types of admixture.

Water reducing admixtures

Water-reducing admixtures (or plasticizers) either reduce the cement requirement for a given strength or increase the workability which is so important for concrete pumping. Reducing the cement content in mixes with a high cement content can not only be commercially beneficial, but could reduce pipeline friction due to the high fines content. When long pipelines are used and particularly when using crushed aggregates and sands, this type of admixture offers considerable benefits. Concrete containing superplasticizers produces what is termed flowing (high workability) concrete and is used extensively for large flat pours such as factory floors. Careful consideration has to be given to the coarse/fines ratio and quality of materials when the mix of superplasticized concrete is designed.

The material is pumpable, but difficulties can arise if there is any significant vertical length of pipeline. A well organized pour of superplasticized concrete with a good mix design and an uninterrupted ready-mixed concrete supply is well worth seeing. The flowing or self levelling concrete looks remarkable when initially placed, and the finishing of the surface using power float finishers starts after only a few hours. The use of superplasticizers in producing flowing concrete is described in a joint C&CA/CAA report.[18]

Water-thickening agents

This type of admixture increases the viscosity of the system which in turn has the effect of reducing pressure bleeding by making the mix more cohesive. Water-thickening agents are used mostly on the leaner mixes but can still be used to advantage in any mix to provide lubrication and so reduce friction.

This type of admixture can be blended with water-reducing agents in the pumping of lightweight concrete (see section 4.6). Kempster[13] describes an experiment where flocculating materials were added in the form of cellulose ethers to unpumpable mixes. Table 12 shows the results when testing the effects of voids in combined aggregate related to cement content as an indication of pumpability. Four out of six of the mixes which hitherto could not be pumped became pumpable when cellulose ethers were added.

Air-entraining agents

Air-entraining admixtures are being specified more and more for exposed concrete — in bridges, roads and airfields. There is no doubt about the effectiveness of this type of admixture which has been well proven in increasing the durability and permeability of concrete against frost.

Adding an air-entraining admixture creates very small air bubbles throughout the concrete. This alters the properties of the concrete both in its plastic and hardened state. As far as concrete pumping is concerned, in the plastic state workability is increased and in an average concrete the presence of air bubbles enables both the water content and the proportion of fine aggregates to be reduced. The concrete is less liable to segregate and bleed. On the face of it, therefore, there appear to be considerable advantages in pumping concrete which has air entrained in it.

However, there are certain precautions which must be taken when placing air-entrained concrete by pump. All pump hiring contractors will need to know details of the mix and particularly the air-entrainment dosage, before starting pumping. If the dosage is too great then problems will arise in pumping. There have been many instances on site when failure to pump the mix has been proved to be caused by too much air in the concrete. The natural reaction of the placing gang is that the pump must be at fault and that it is not good enough for the job. It should be remembered that air is

*Table 12. Results of experiment to test effect of voids in combined aggregate related to cement content as an indication of pumpability**

Cement content by volume: %	Void content by volume: %			
	17	20	25	28
15	Pumped	Pump blocked during experiment	Did not pump†	Did not pump
20	Pumped	Pumped	Did not pump†	Did not pump
25	Pumped	Pumped	Pumped	Did not pump†
30	Pumped with difficulty	Pumped	Pumped	Did not pump†

* Sand content was 35% of combined aggregate in all mixes; water content was adjusted to give 75 mm slump.
† Became pumpable when cellulose ethers were added.

Table 13. Air contents recommended by BS 8110

Nominal maximum aggregate size: mm	Total target air content: %
10	7
20	5
40	4

compressible and part of each piston stroke is being uselessly expended in doing just that — forming a cushion of air which can inhibit the movement of concrete along the pipeline. Therefore if the dosage of an air-entrainment admixture is too high there could be problems in pumping. Air contents as recommended by BS 8110[19] are given in Table 13.

Pump hirers would probably quote guideline figures similar to the ones given in Table 13, plus or minus 1·5%. Off-the-boom pumping is usually trouble-free with less than 6% air content using 125 mm dia. pipeline. The longer the pipeline the greater the pressure and this could add to the problem. But a sensible approach to the job should overcome such problems. Consideration should be given to increasing the diameter of the pipeline, keeping the number of bends to a minimum and perhaps adding a plasticizer, if there is concern over the dosage.

One further note of caution when pumping air-entrained concrete is that the volume of the air is less at the discharge end than at the pump as some air is lost when pumping.[20,21] This can be typically 20% or more of the volume of air entrained (Fig. 64). It is usual to allow for this by increasing the amount of air, but if this is overcompensated for, the concrete may fail to be pumped. Generally, adding no more than 2% to the specified dosage when pumping over about 50 m should normally deal with this problem without creating any pumping problems. It is common practice nowadays to use air meters on site for testing the air content at the placing position to BS 1881 Part 106, 1983.

Special care should be taken when designing a mix which incorporates air-entraining admixtures. Fine ground cements such as Rapid Hardening, PFA, and GGBFS, are all finer than OPC and it is necessary to increase the dosage to obtain the required amount of air entrained into the mix. This can be up to twice the normal dose. More water may also increase the amount of air entrained. Within limits, the longer the mixing time or the smaller the batch, the greater the amount of air entrained. It is recommended that trial mixes be carried out when using air-entraining admixtures.

Accelerators and retarders

Although mixes containing accelerators will pump normally, the fact that the concrete will go off more quickly must be borne in mind. The obvious concern is that during

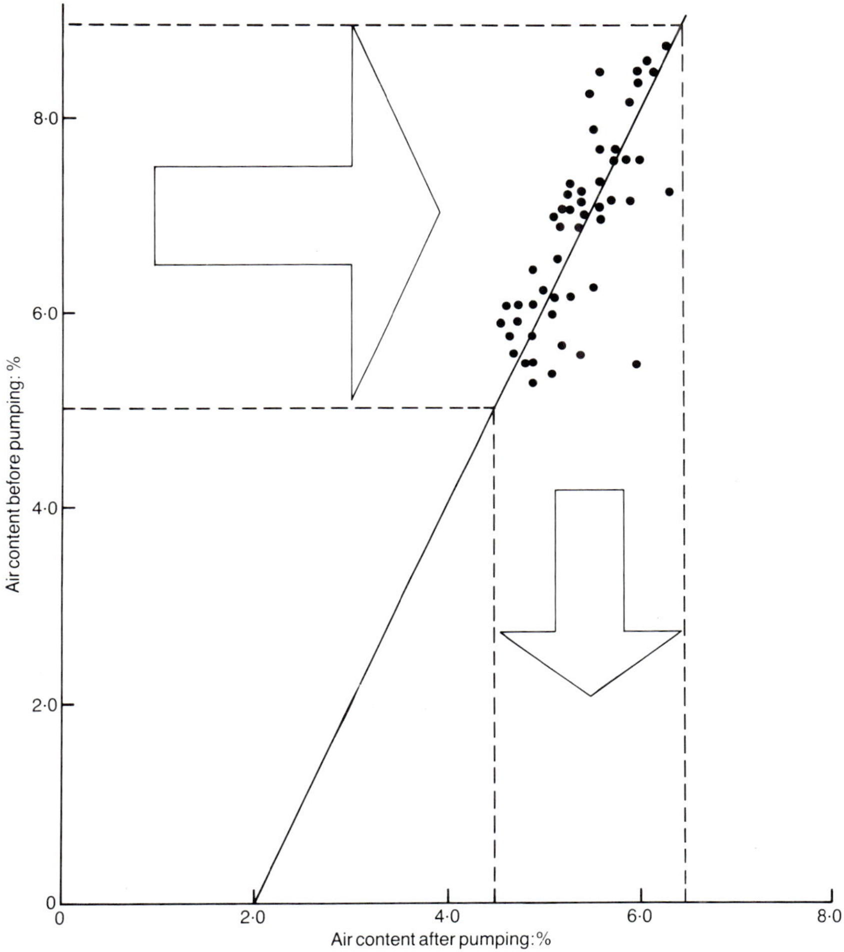

Fig. 64. Air content in fresh concrete measured before and after pumping

a stoppage, which may be for a variety of reasons, e.g. erratic concrete deliveries, slow placing methods, pump breakdown (however rare this might be), the concrete will go off quicker in the boom or pipeline than a mix without an accelerator. Clearing blockages is much more difficult. At the first sign of delay, the pump operator has to start thinking of washing out the pump and the complete delivery line.

Retarded mixes can be pumped readily and indeed are used extensively, particularly on large pours such as thick slabs, or for reasons generally to offset the effects of high

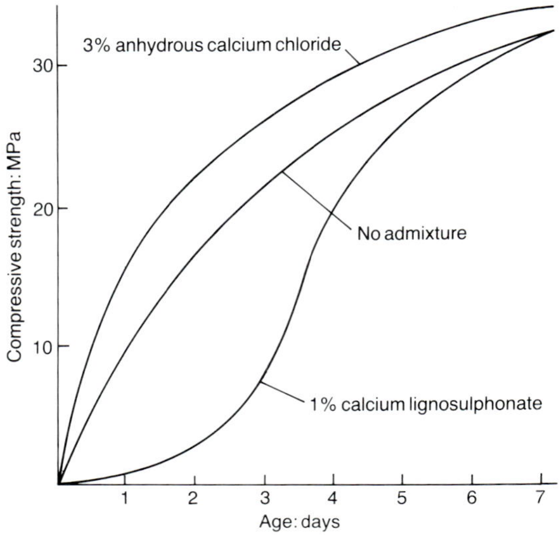

Fig. 65. Effect of accelerator and retarder on strength development of a typical 1:6 concrete at 18°C

temperatures. They are very useful for long distance pumping. The use of a retarder in the pumped concrete for a thick railway bridge deck over the Rhine at Dusseldorf was observed by the Lancaster University research team in their concrete pumping study.[1] Here a retarder was used for the concrete in the bottom half of the slab and the concrete in the top half was placed without a retarder. This was to avoid cold jointing and the whole slab set at roughly about the same time. Retarders can also be used effectively in slipform work to obtain structural continuity between lifts.

The Lancaster University study[1] also instanced a pour in West Germany in mid-summer with very high temperatures, when the use of a retarder could have prevented concrete setting on a floor slab pour before screeding could be completed. Remedial action eventually retrieved the situation, but there was a costly loss of material as well as a serious delay to the work.

Fletcher[22] deals at some length with admixtures and Fig. 65 shows the variations in strength development when using a retarder or accelerator.

Other types of admixtures for use in pumpable concrete

Finely divided material admixtures can aid pumpability. PFA (or fly ash) is perhaps more widely used than any other admixture throughout the world. The use of microsilica (or silica fume) is becoming more widespread. These are described more fully in section 4.5.

Blends of admixtures, pumping aids, and underwater concrete admixtures are now available. All should be used with caution, in accordance with the manufacturers' instructions. Some underwater concrete admixtures need higher line pressures which are undesirable.

Steel, glass fibre or polypropylene fibres can be added to the mix which is readily pumpable, but extreme care must be exercised to prevent blockages and damage to the pump through increased wear.

Information on admixtures

A comprehensive list of admixtures including those already mentioned can be obtained from the CAA.[23] The various types of admixture are described, and details of the ingredients, dosages and other useful information are also provided under each manufacturer or trade name. There is also a section on aids to pumpability.

As with any type of admixture, particular care is required to obtain the correct dosage for a specific application or circumstance and the advice of the manufacturers should always be sought. After all, they have spent a great deal of time and money on extensive development and testing of their products. In any case, the prudent contractor will ensure that successful trial mixes have been completed before using any type of admixture. Up-to-date information on developing trends with admixtures and how best to take advantage of these materials can be obtained from *Cement admixtures: uses and applications.*[24]

4.5. Cement replacement in pumpable mixes

PFA to BS 3892[25] and GGBFS to BS 6699[26] are widely used as cement replacement for many classes of concrete including pumpable mixes. They are recognized for use in structural concrete to BS 8110. There has been much research into the use of these materials and it is generally accepted that their controlled use not only cuts the cost of the concrete but also improves its durability. There have been some misgivings as to early strength gains and long-term effects from the use of these materials, but a report from the C&CA published in December 1986[27] stated that 'perfectly satisfactory concrete can be made with composite cements containing pulverized fuel ash or granulated blast-furnace slag.' A proviso was made as to adequate quality control and care during the early curing stages of the concrete. This followed an earlier agreement by the Department of Transport in its 'brown book',[28] which was published the month before, that cement substitutes could be used for structural concrete.

PFA

PFA is a by-product of the pulverized coal which is burnt at most power stations and it has been used for some years for load bearing fill for roads and embankments and for grouting. It is a fine powder with a fineness about equal to or finer than Portland cement and it consists principally of oxides of silicon, aluminium and iron.

The producers of the material claim that the main properties of fineness, pozzolanic activity, which is cementious behaviour in the presence of free lime, rounded particles and reduced water demand, have made it suitable for all classes of concrete including pumped concrete. It is claimed that the many benefits offered by the correct use of this material incude improved resistance to sulphates and alkali-aggregate reaction. The concrete has higher long-term strength and reduced permeability. The disadvantages of slower early strength gain perhaps diminish in importance when the proven benefits in durability are considered.

Core samples taken from the PFA concrete placed 25 years ago at High Marnham power station, Notts, showed that the strength increase from one year to 25 years was twice as much for PFA concrete than for Portland cement concrete. Lower porosity and virtually no carbonization were other advantages: these are described by Carbrera and Wooley.[29]

Pumping PFA concrete

Manufacturers claim that a pumpable mix with this type of additive can prevent or reduce pressure bleeding without increasing pipeline friction to the extent that would occur if additional cement was used. Most pump owners and operators are happy to pump a PFA mix and the majority say that they prefer this to an OPC concrete. Using PFA as partial cement replacement increases the volume of cementitious material in the mix and, provided that this is controlled, i.e. that the quantity does not exceed that described in section 4.2, pumpability is improved. The particles are rounded and this is an aid to pumping. Reflecting on two of the basic design requirements for a pumpable mix, namely correct grading of aggregates and sufficient fine fines to fill the voids, it would seem that the use of PFA as part of the cementitious material can be beneficial. It has to be stressed that careful mix design is essential and only top quality material to the appropriate British Standard should be used.

As cement is the most expensive material in a concrete mix, any reduction in the quantity used, provided that the end product (the concrete in its finished state) is up to specified requirements, must be a favourable proposition.

Examples of the use of PFA

At Portrack sewage treatment works, Northumbria, part of the 16 000 m^3 of concrete which contained 30% PFA replacement was placed by pump, in 1982–83. Factors to be considered were ground sulphates, exposure to sewage, frequent wetting and drying due to changes in levels of flow and severe weather conditions in winter. All such conditions favoured the use of PFA as a partial cement replacement, but it was mainly on economic grounds that permission was given to use PFA. There were considerable savings in costs.

The mix details are given in Table 14. An air-entraining agent was used and aggregates were single size. 28 day strengths of 40 and 50 N/mm^2 were obtained. There was no delay in formwork stripping times, which was an initial concern.

At Drax power station, Yorkshire, a total of 160 000 m^3 of PFA concrete was used in the main foundations and ancillary structures, as described by Wooley.[30] The details of the pumped concrete mix, which was a 30/20 mix, are given in Table 15. More study and discussion may need to be carried out if full advantage of PFA as a cement replacement is to

Table 14. Mix details using PFA replacement

Material	Normal concrete	Pumped concrete
20 mm aggregate: kg	820	720
10 mm aggregate: kg	360	360
Sand: kg	610	660
OPC: kg	280	290
PFA: kg	120	125
Water: l	170	185

Table 15. Pumped concrete mix at Drax

20 mm aggregate	670 kg
10 mm aggregate	445 kg
Zone 3 sand	745 kg
OPC	240 kg
PFA	150 kg
Plasticizer	1·365 l
Slump	110 ± 25 mm
Water/cement ratio	0·46
7 day strength	32·4 N/mm^2
28 day strength	49·0 N/mm^2

be taken. Or perhaps it is a case of more education on the subject being required.

GGBFS

GGBFS is an ironworks blast-furnace by-product in the form of finely ground slag containing coal and iron. It has excellent cementitious properties and its chemical composition is similar to Portland cement. In the manufacture of GGBFS, the slag emerging from the blast furnace at a temperature of about 1500°C is quenched in water, which produces glassy granules. These granules are later dried and then ground to a fine powder. Although GGBFS is almost classed as a cement in its own right, GGBFS cannot be used on its own and requires activation to develop its cementitious properties. Thus it is used with Portland cement and the combination is known as Portland blast-furnace cement (PBFC). PBFC is manufactured to BS 146 and BS 4246.[31,32]

GGBFS has been used for many decades in most parts of the world. In the UK, nearly all the ready-mixed concrete plants have silos for this material, which is available under several brand names. It is now recognized throughout the construction industry that, if used in conjunction with Portland cement, GGBFS will provide a highly satisfactory concrete of high strength and durability. It is used in all types of structural and non-structural concrete. Manufacturers claim with some justification that the inclusion of GGBFS improves many of the characteristics of concrete made with OPC.

The benefits of GGBFS which particularly affect concrete pumping include workability. Excellent flow characteristics are clearly evident and blockage problems are minimized. As the hydration process is slower when using PBFC, the concrete remains workable for longer and this can be advantageous on larger pours.

Heat of hydration is considerably less when using PBFC. Blundell and Browne[33] state that a 50% cement replacement using GGBFS will result in a 40% reduction in maximum temperature achieved. There is no doubt that thermal cracking is reduced if this material is incorporated in the mix, particularly where large deep pours (which are particularly suitable for placing by pump) are concerned or in high cement content mixes. Fig. 66 shows the effect of slag on the temperature rise due to heat of hydration.

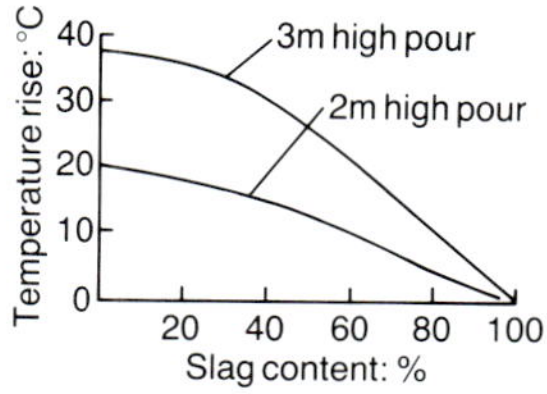

Fig. 66. Effect of slag on heat of hydration temperature (360 kg/m³ slag and cement)

Resistance to sulphates is another benefit from the use of PBFC. In most conditions using 70% PBFC and 30% OPC can be used as an alternative to sulphate resisting cement. BRE Digest 250[34] gives full details and recommendations on this subject. The other important advantage to be obtained from the use of this material is in protection against the alkali-aggregate reaction. It is claimed that a simple and effective measure against this is to design a mix with 50% or more GGBFS replacement. CS technical report 30 describes procedures for minimizing this risk.[35]

Strength development using GGBFS

Blast-furnace slag cements create a reduction in early age strength gains, which may be a disadvantage in certain cases. This reduction is due to the slower rate of hydration of the slag component of the mix, which also leads to the reduction in temperature. Blundell and Browne discuss this in reference 33. Fig. 67 shows the effect of using GGBFS in different proportions on compressive strength.

Microsilica

Silica fume, or microsilica as it is now more commonly called, is another form of cementitious material which is used as a cement replacement and as a mineral additive in concrete. Microsilica concrete was the subject of two CS current practice sheets[36,37] which comprehensively cover the physical and chemical characteristics of the material, as well as types and grades, effects on concrete properties and particular applications. These include its use in lightweight concrete, grouts and mortars, repair materials, underwater concrete, shotcrete and fibre reinforced concrete. Microsilica concrete is readily pumpable as it complies with all the requirements for a pumpable mix.

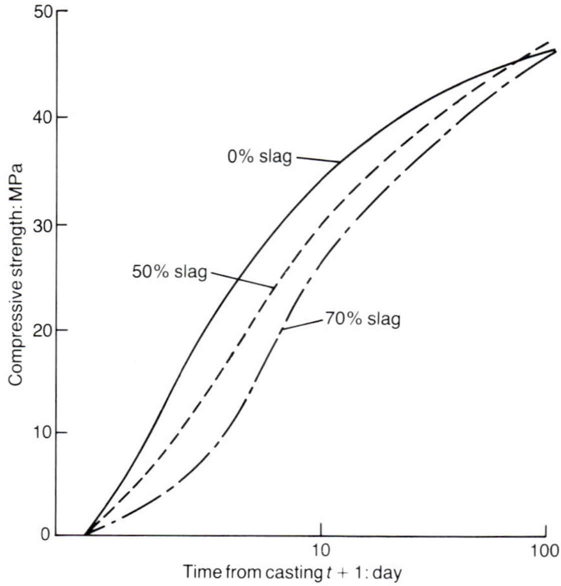

Fig. 67. Effect of slag on the strength of the concrete (370 kg/m³ SRPC)

Microsilica consists of minute spheres of silicon dioxide (SiO_2), which is a by-product of the manufacture of ferro-silicon. The average particle size is 100 times smaller than a cement grain. The manufacturers claim that when microsilica is included in a concrete mix there is a physical effect and a chemical effect.

The physical effect is that the material provides the necessary fines below 300 μm to assist the lubrication of the mix. Once dispersed, the microspheres counteract bleeding and segregation. The material imparts a quasi-thixotropic prop-

Table 16. Aggregate grading for mixes containing microsilica

BS sieve mesh size: mm	Percentage passing	
	20 mm	10 mm
20	100	–
10	58–75	–
5	40–57	100
2·36	36–55	78–82
1·18	26–44	35–53
0·6	17–34	22–44
0·3	12–22	10–42
0·15	7–10	3–18

Table 17. Cement content guidelines using microsilica

Mix	OPC content: kg/m^3	Microsilica content: kg/m^3	Equivalent cementitious material: kg/m^3
C20	200	18	$200 + 18 \times 5 = 290$
C25	220	22	$220 + 22 \times 5 = 330$
C30	250	25	$250 + 25 \times 5 = 375$
C40	300	30	$300 + 30 \times 5 = 450$

erty to the mix and facilitates pumping without the need for additional sand, water or cement.

The chemical effect is that the microsilica converts calcium hydroxide released during the cement hydration process to calcium silicate hydrates, resulting in a higher strength concrete, less shrinkage, reduced permeability and greater durability.[38]

The only microsilica available in the UK which is covered by Agrément Certificate 85/1568[39] is manufactured and distributed by Elkem Chemicals who suggest the guidelines given in Tables 16 and 17 for pumpable mix designs which include microsilica. The slump should be 75 ± 25 mm and the coarse/fines ratio approximately 3:2. The bulk density of microsilica is about 275 kg/m^3 and OPC about 1440 kg/m^3. Therefore 1 kg of microsilica is equivalent in volume to 5 kg of OPC. Thus in the mixes indicated in Table 17, if the microsilica content is multiplied by five and the resultant figure added to the OPC content, an equivalent figure for the total cementitious material can be calculated.

Reports on the cementing efficiency of microsilica in terms of producing equal compressive strength to OPC concrete using gravel and limestone aggregates,[40,41] conclude that the cementing efficiency factor (K) was as given in Table

Table 18. Cementing efficiency factor K

Use of microsilica	Gravel aggregate factor K	Limestone aggregate factor K
Using microsilica in proprietary form	3–4	4–5
Using microsilica in slurry form plus calcium lignosulphonate plasticizer	6	7–8
Using microsilica in slurry form plus melamine formaldehyde superplasticizer	8–9	6–7

18. It was clear that the factor K is dependent on the efficiency of the dispersant. It was also found that despite a mean increase in water/cement ratio of 0·11 (gravel) and 0·17 (limestone), microsilica concretes achieve equal strengths at the same workability compared with OPC concretes.

4.6. Pumping lightweight concrete

With the publication of a new guide to the use of lightweight concrete (LWC) in February 1988,[42] the UK construction industry had an authoritative document on the structural use of this material, although more and more use has been made of it in recent years for floor slabs and roof screeds. The advantages of the material include reduced loading, better fire resistant properties, better durability, savings in maintenance costs and improved heat insulation.

Manufacturers of the lightweight aggregate may find it a problem to persuade some engineers to specify their product, but there is no problem in pumping it. Reluctance to specify LWC may stem partly from misconceived ideas concerning cost. However, a CS paper on LWC[43] in 1983 stated that in spite of higher materials costs, LWC structures work out marginally cheaper when proper cost comparisons are made which take into consideration the advantages which have already been mentioned.

The rounded nature of the lightweight aggregates is, of course, a distinct advantage for the pump mix designer, as this will reduce pipe friction and consequently a lower pumping pressure is required, all of which would point to a lower probability of bleeding and segregation. The problem of water absorption by the lightweight aggregate, such as foamed slab or Lytag, can be overcome by vacuum soaking the material, using a low pressure pump, or adding an admixture. The most practical solution, at least in the UK, has been found to be the use of a pumping admixture.

Table 19. Mix design using lightweight concrete

Material	Quantity/m^3
Lytag pump grade (SSD)	520 kg (0·65 m^3)
Zone M natural sand (SSD)	810 kg
OPC	325 kg
EMSAC 1000 WR	65 kg
Water	195 l

Fig. 68. Spread test

Mix designs will vary, but a typical mix using Lytag aggregate and a microsilica based plasticizer/water thickener admixture such as EMSAC 1000WR, would be as given in Table 19. A continuous grading (as with all pump mixes) is required between the sand and coarse aggregate. Grading of the natural sand is very important and is normally about 20% passing the 300 μm sieve. A figure of about 10% would be too low. The plasticizer enables the admixture to disperse readily through the mix although the thickening effect takes a little while to appear.

Flow tests are carried out using a flow table as described in BS 1881: Part 105, 1984. The flow table is a 600 mm × 600 mm board with a hinge on one side, and a lifting handle on the opposite side to the hinge. A specified amount of concrete is poured into a cone, and the cone is then emptied on to the flow table. The table is raised, by the handle, 50 mm and dropped 15 times. The flow or spread is defined as the mean of two diameters measured after the table has been raised and dropped 15 times (Fig. 68). Slump testing is not considered to be a reliable measure of workability for the flowing concrete produced for pumping, as a slump test would produce a 'collapse' result. The practical pump-hire operators maintain that as a guide for the initial load of concrete, a flow reading of between 600 and 650 mm will ensure that the material will be suitable for pumping.

It is necessary when batching to assess the moisture content of the sand and the lightweight aggregate and to make an allowance for the additional absorption

Water required = effective water content
− moisture in the sand
+ water for the additional absorption of the lightweight aggregate

An example of this is shown in Pozzolanic Lytag technical bulletin 2.[44] Lytag aggregate will absorb water at approximately 15% by dry weight when subjected to pressure from concrete pumps. An example of mix adjustment using Lytag aggregate is given in Table 20. The procedure to be adopted is as follows

(*a*) measure moisture content of sand and Lytag; the Lytag should be partially crushed
(*b*) adjust dry weights accordingly to give batch weights
(*c*) reduce amount of effective water by the total weight of water in the aggregates
(*d*) add the 15% absorption allowance to the result of (*c*) giving the mixing water required.

In the example given in Table 20, for grade 30 concrete, the

Table 20. Mix adjustment using Lytag aggregate

Material	Dry weight: kg	Corrected batch weight/m³: kg
OPC	410	410
Sand	730 × 1·05	767
Lytag pump grade	536 × 1·06	568
Effective water		210
Less moisture content of		
Sand	730 × 0·05	−37
Lytag	536 × 0·06	−32
Total		141
Add aggregate absorption allowance	536 × 0·15	+80
Total		221
Mixing water required		221 l/m³

Table 21. Typical mix design per cubic metre of compacted concrete at average flow

Characteristic 28 day strength: N/mm^2	OPC to BS 12: kg	Suitable sand to BS 882 (dry): kg	Lytag pump grade to BS 3797: m^2	Admixture suggested dose: kg	Effective water: kg	Workability flow range to BS 1881: cm
20	330	800	0·65	1·0	210	50–60
25	360	770	0·65	1·0	210	50–60
30	410	730	0·65	1·0	210	50–60
35	470	690	0·65	1·0	210	50–60
40	550	620	0·65	1·0	210	50–60

moisture content of sand and Lytag pumping grade are 5% and 6%, respectively. The dry weight of Lytag aggregate is 0·65 × 825 kg = 536 kg.

Table 21, from the Lytag technical bulletin, illustrates a typical mix design. The manufacturers stress the importance of even grading of the sand and quote a Lytag bulk density, as a guide only, of 825 kg/m^3. Regional variations should be checked.

4.7. QA and the ready-mixed concrete industry

Ready-mixed concrete suppliers, which operate over 1200 certified plants are QSRMC[45] members. QA is now a widely accepted part of building and civil engineering and the ICE has strong links with the BSI Committee on Quality Assurance.

The QSRMC is based on BS 5750[46] and is given considerable coverage in the *Manual of ready-mixed concrete*[47] in which it is made clear that QA covers the whole production process — not only the material, but also plant and equipment. The scheme has the full backing of the two ready-mixed concrete associations, the BACMI and the BRMCA. Evidence of the practicalities and success of the scheme were observed by the research team in the recent study on ready-mixed concrete[48] in the East Midlands area of the UK, where 50 plants were visited. The value of such a scheme was clearly evident.

However, the possession of a QA certificate is not the key to commercial success, but the company involved in quality assurance will eventually benefit as QA will become part of the overall strategic outlook of the management. The principles of such a scheme can be broadly described as the implementation of documentation to control the quality of the production or service. It provides a disciplined approach and can be reviewed by management from time to time as may be necessary.

The concrete pump manufacturers have their own quality control systems and many of the pump hire companies have written procedures which include the setting up, operation and maintenance of the pump, laying of the pipeline and washing the pipeline and pump out. Safety is also emphasized, not only with regard to the pump unit and the operator, but also to the general public. Correct operation of the appropriate pumping equipment in good condition for a specific job will ensure that there is no loss in quality of the concrete from a ready-mixed or site batched source. Such firms are in effect operating their own quality control.

Certification of pump operators will be required by law in the UK in the near future and it may not be very long before there is a recognized quality scheme for concrete pumping.

5

Special applications

Concrete pumping is used for many types of pours. Four special applications to which a concrete pump is particularly suited (and in some cases is a requirement) will be described here, namely, concrete pumping in tunnels, underwater concreting, filling of flexible forms and mattresses and continuous flight auger piling. There are distinct advantages using a concrete pump for each of these applications.

5.1. Concrete pumping in tunnels

The size of the tunnel, the working space and the method of construction adopted by the contractors will often dictate the type of concrete pump and equipment which must be used. For example, the tunnel may be so small that it is not possible to accommodate the concrete pumping unit, or working space may dictate that all the concrete pumping plant is positioned at the tunnel entrance and only the delivery pipeline is inside. Larger tunnels may be able to take mobile concrete pumps with booms and there may be access for ready-mixed concrete trucks to supply concrete directly to the pumps.

It may be possible for concrete materials to be brought up to the pump at the working face by means of a so-called tunnel train; the dry materials are mixed on the spot and the concrete supplied directly to the pump hopper. There are many variations and applications in this type of specialized construction work. Most standard mobile or trailer pumps are used on tunnel work; the types used depend on the circumstances and the particular application. The pumping range of many of the pumps is such that they are ideally suited to conveying concrete over long distances, something which is often required in tunnel construction. Normally, production requirements for concrete in tunnel construction are not high. The main problems, at least as far as the concrete placing and associated work are concerned, are in the formwork, reinforcing steel fixing and pumping the concrete up to the pour position. Consequently, many of the lower output pumps which can be positioned inside a tunnel are very suitable.

On the other hand, if the tunnel is large enough, mobile pumps are often used. These pumps have generally the same pumping units as the portable pumps, but the booms are specifically designed for use in tunnels. The boom design is aimed at achieving the lowest unfolding height and maximum flexibility in swivelling the end section (Fig. 69). Accuracy in positioning the end delivery section is guaranteed without having to use a flexible end hose, which would be a cumbersome and difficult operation in a tunnel.

Fig. 69. Performance diagram for the adjustable boom tip of the BP350 tunnel pump; dimensions in mm (courtesy F. W. Schwing/Burlington Engineers)

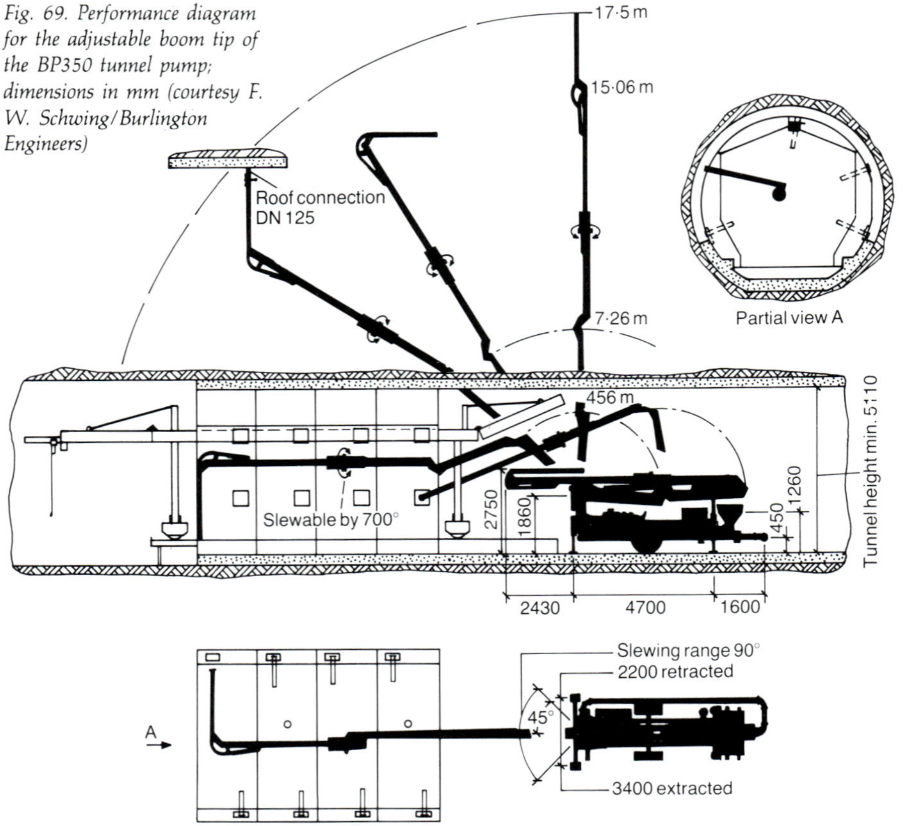

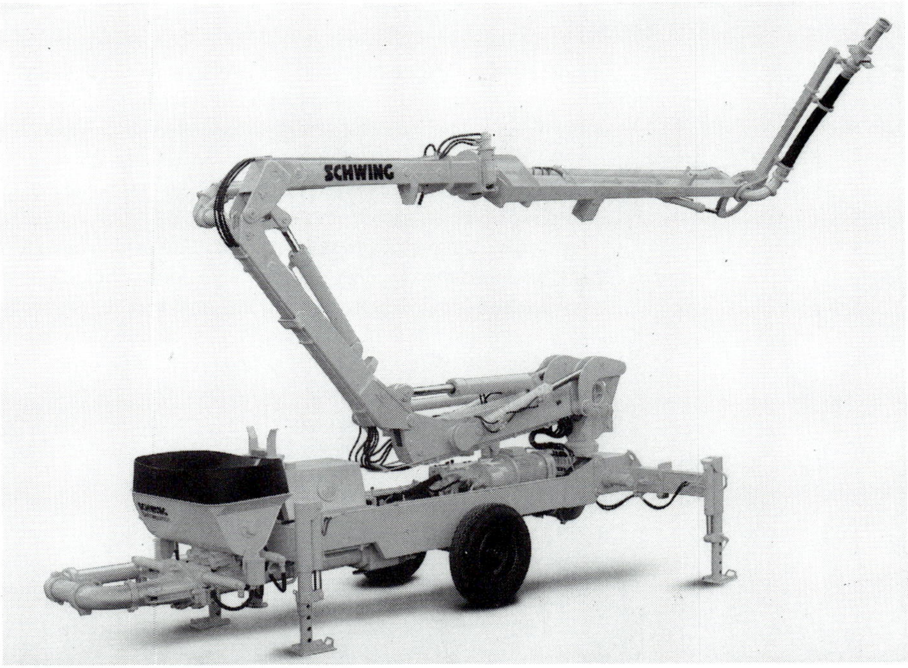

Portable pumps fitted with a placing boom can be used in relatively small tunnels. The articulated tip is hydraulically operated and can be rotated through 360°. For connection to permanently installed sockets in the formwork, a cardan-type coupling is used (Fig. 70).

Fig. 70. Schwing BP350TVDM trailer tunnel pump with special tunnel boom (courtesy F. W. Schwing/ Burlington Engineers)

Tunnel trains

On large tunnel projects when long distances are involved inside the tunnel, the so-called tunnel concreting train is frequently used. This consists principally of four items, usually all mounted on rails (Fig. 71). Materials (aggregates and cement) can be premixed or brought up separately from

Fig. 71. Typical concreting tunnel train (courtesy F. W. Schwing/Burlington Engineers)

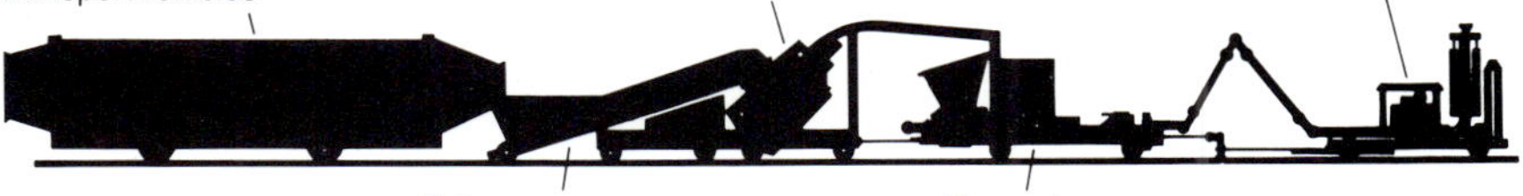

outside the tunnel. The materials are loaded on to the conveyor which feeds the mixer. The conveyor can be a smooth belt or screw type. Mixers are usually single shafted, batching is by weight and mixing capacity up to 30 m^3/h. From the mixing unit the now plastic concrete is fed to the pump. The distributing plant used will depend on specific requirements and may be a simple pipeline or a special tunnel distribution boom.

Using a tunnel train similar to the one already described, between 6 m and 13 m of 25 mm thick full circle concrete lining was placed every day in the Freudenstein tunnel in West Germany. A Putzmeister TCT15 tunnel concreting train was used (Fig. 72): the dry materials were pre-mixed outside the tunnel at a batching station and brought up in a single silo rail car. This container car was large enough to carry sufficient material for concreting 60 m of tunnel lining.

A similar type of tunnel train was used for concrete lining work to galleries on the Kielder Water scheme in Northumbria, England, where 28 km of 250 mm thick lining to a 3·5 m dia. tunnel was completed. The nature of the job made it possible to use a circular formwork length of 48 m in 6 m length sections. Working 24 hours a day with a formwork stripping time of ten hours (6 m long sections were stripped and refixed at the leading end of the tunnel), resulted in 100 m/day of tunnel being completed. The concrete output was in excess of 300 m^3/day. These are very high perfor-

Fig. 72. Tunnel train (courtesy Putzmeister Ltd)

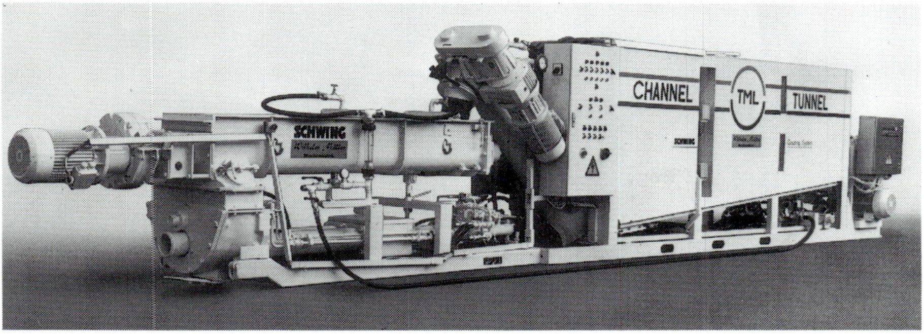

Fig. 73. Grout pump/mixer (courtesy F. W. Schwing/ Burlington Engineers)

mance figures. Schwing high pressure pumps rated at 27 m^3/h with a maximum placing pressure of more than 80 bar were used. The 125 mm dia. pipeline was laid on the top of the formwork and was restrained there by steel supports. The whole of the pipeline, which was approximately 50 m long, moved as part of the complete tunnel concreting train.

A modern type of tunnel grout pump with a mixer attachment is shown in Fig. 73. This mortar pump or injection station, which was used on the Channel Tunnel for grouting behind tunnel segments, has an output of up to 10 m^3/h. It incorporates a Müller twin shaft mixer and storage bin. Grouting materials are brought to the pump position by separate mobile feed bins. The system can be remote controlled from the tunnel heading working position.

5.2. Underwater concrete pumping

The use of a concrete pump is widely accepted as a suitable option for placing concrete underwater. The fact that the material is to be placed through water and that it will displace water as it is cast into place clearly defines the problem. Overcoming the problem of placing concrete underwater is not a difficult matter if correct procedures are carried out. There are, of course, other problems which may or may not arise, such as the effects of tidal or other water movements on the newly placed concrete, but these effects can be dealt with in other ways.

It is appropriate first to describe a simple tremie operation, as the same principle is applied when using a pump for underwater concreting operations. The word 'tremie' is derived from the French word for hopper and the system has been used since the beginning of this century. Fig. 74 illustrates the operation. A hopper with pipe connections

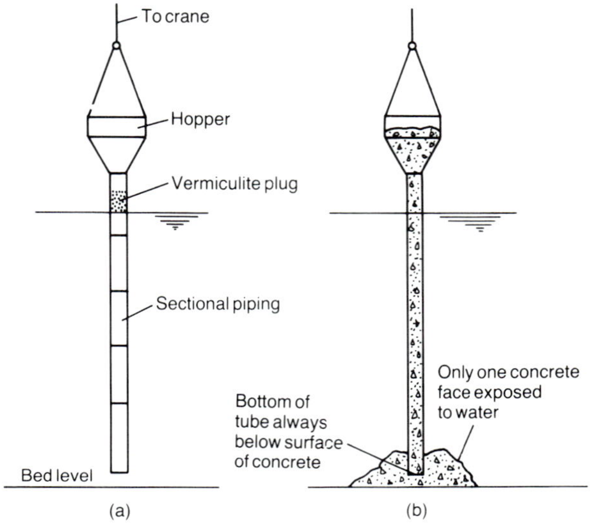

Fig. 74. Underwater concreting by tremie: (a) before concreting; (b) during concreting

attached is suspended from a crane and lowered into the water until it rests on the bottom of the pour. The water will rise up the pipe to reach the same level as the surrounding water. To protect the concrete as it emerges from the end of the pipe, a vermiculite plug which is roughly 1·5–2 times the pipe diameter long is inserted. Sponge balls are sometimes used for the same purpose. Concrete is poured into the pipe. When the pipe is filled the end of the pipe is raised slightly off the bottom, sufficient for the concrete to emerge, pushing the vermiculite out in front of it. This prevents the water and concrete from mixing. The pipe must be kept topped up and can be moved laterally or vertically as required. It should be remembered that the end of the pipe must remain in the concrete during the whole of the concreting operation.

When using a pump for underwater concreting, adhering to the basic prinicples of tremied concrete is essential. A concrete pump offers advantages in that the method is neater and less cumbersome than a suspended hopper/pipe arrangement which has to be kept filled by other means. Also, when a pump is used, the spread of the material is greater when the pipe is at any one position and there is better compaction because of the higher placing pressure of the pump. It should be noted that it is not normal to use vibratory methods for compacting underwater concrete.

Operation

Both trailer/static pumps and mobile pumps are used for placing concrete underwater and it is possible to use flexible hose ends as well as steel. The pump delivery line is in fact taking the place of the tremie pipe. Steel ends are used vertically, but the last 2–3 m of a flexible end hose used in certain conditions should be laid horizontally. If this length of pipe is buried in the concrete it takes the surge pulsations out of the end section, which otherwise could prove troublesome. If the pipe is moved laterally, the method is very similar to that of the hydrovalve system which was developed in Holland. The hydrovalve system entails moving the valve assembly in the direction of the pour, and the surface of the concrete is levelled simultaneously (Fig. 75).

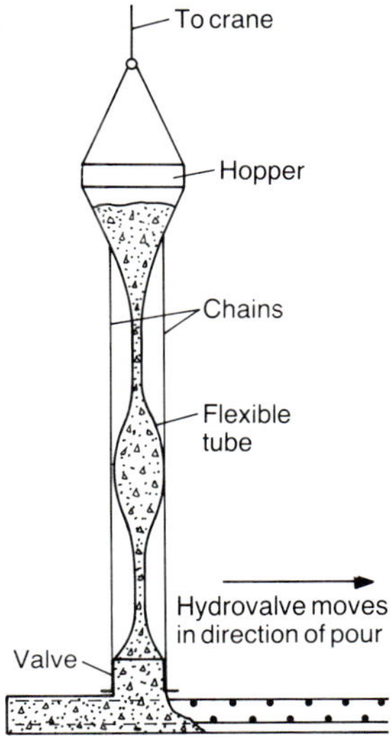

Fig. 75. Principle of the hydrovalve

The normal operating and safety procedures used in conventional concrete pumping (chapter 12) will apply to underwater work. Gravity feed precautions, e.g. when grouting the pipeline, provision of shut-off valves, non-return valves for greater depths, etc., are obviously important.

If blockages occur when pumping concrete underwater they can be very serious. Detecting and clearing blockages is described in chapter 8, but clearing a blockage when part of the pipeline is underwater presents different problems. Pumping back a full pipeline of concrete in the event of a blockage could also pump back water, with disastrous results. If a blockage occurs, a check should be made underwater first to see if there is evidence of any problem there, e.g. kinking, if a flexible hose is being used. The line should then be uncoupled above water level and the pump reversed. More often than not, this will clear the blockage. If not, the blockage will be located in the length of pipeline above water or in the pump valve assembly and must be cleared by hand. If the blockage is underwater, the whole length of the pipeline has to be brought up to ground level and the blockage removed manually.

Special precautions are usually taken to avoid blockages in underwater concrete pumping. To test the concrete for pumpability, the first load is usually recycled into the pump hopper. In some instances it is considered prudent to lay a second pipeline so that if there is a blockage which cannot be cleared immediately, the supply can be transferred to this line while the first line is cleared. In addition, if it is imperative to maintain continuity of placing, a stand-by pump is kept on site.

Pipe joints should always be the rapid fit type with raised handles and safety pins, as these are the most suitable for a diver, who may be wearing gloves, to deal with in poor visibility. Securing any length of pipeline suspended in water by rope is sensible, as in the event of a joint failure, recovery is easier and injury to a diver could be prevented.

As with all tremie work it is vitally important to be aware of the exact position and depth of the pipe end at all times.

Concrete mix for underwater concreting

The criteria for concrete pumped underwater are similar to concrete pumped in other applications, with the emphasis on cohesive qualities and workability. Generally the parameters for mix design described in chapter 4 should be met. High workability is necessary for self-compaction and this can usually be attained without any loss in strength if proper mix design procedures are followed.

Aggregates should preferably be rounded, especially the coarse aggregates, to increase the flowability. The mix design can include admixtures to satisfy specification requirements. On the Thames Barrier foundations, a retarder was successfully used on the massive pours there.

Non-dispersible concrete

Non-dispersible concrete has been used increasingly for underwater concreting in recent years and although the concept is still relatively new, much research and development has been carried out and is continuing to be carried out. As the name implies, non-dispersible concrete can fall through water freely without the concrete segregating or the cement washing out. It is highly cohesive and water repellent. When used with a suitable plasticizer it may be self-compacting and thus ideal for distribution and levelling underwater. It can be pumped easily and tremie procedures need not apply. Free-fall techniques may be used, if the delivery end of the pipeline is kept at least 1 m below the surface of the water.

To create non-dispersible properties, a microsilica based admixture such as Hydrobond 870 (Elkem Chemicals Ltd) may be added to the standard pumpable mixes as a liquid at the ready-mixed concrete plant or on site. These mixes should have a cement content of 300–400 kg/m^3 and a coarse/fines ratio of 3:2. At least one minute of additional mixing time is required to disperse the admixture thorough-

Fig. 76. Underwater concreting to harbour entrance (courtesy Rapid Concrete Placing Ltd)

ly. Other thickening agents may be used to create the same non-dispersible effect, but certain types may cause problems when cleaning plant after completion of the pour. This is due to the extreme cohesiveness of the concrete. High pressure washing facilities may be necessary.

Strengths of up to 70 N/mm^2 can be obtained readily with non-dispersible concrete and although there is little knowledge yet of long-term results, there appears no reason why there should be any concern over this.

Placing non-dispersible concrete underwater can be carried out at higher production rates than when placing conventional mixes, as the constraints of carrying out a tremie operation do not apply. Also resistance to erosion when in its plastic state is another advantage of non-dispersible concrete, as succeeding lifts or layers require only minimal preparation. Non-dispersible concrete is not limited to use underwater and can be placed by any of the other recognized methods of placing concrete above water.

Finishing level requirements will vary from job to job. It is possible for divers to use rolled steel sections to screed

Fig. 77. Underwater concreting marina construction: pump operator has telephone communication with diver (courtesy Pochin Group plc)

underwater concrete. Accuracies of $\pm$ 10 mm are claimed. In most cases examination of the concrete surface will show that the concrete was self levelling and only the top 50–60 mm has been affected by the water. This can be dressed off 24 hours after placing. Concrete placing underwater does not have to suffer because of changing weather conditions above ground. Water at a reasonable constant temperature provides good conditions for curing. In some respects, there are therefore advantages to placing concrete underwater. There should be no problems in obtaining a good mix design to fit specification requirements. All that remains is to ensure that the placing method does not damage the concrete by allowing any loss of fines (sand and cement) and the finished result can be entirely satisfactory.

It is also possible to pump concrete into rigid formwork or various flexible forms underwater (see section 5.3). Rigid formwork should be steel to prevent flotation and should be as simple as possible, bearing in mind that divers will be used to fix and strike the shuttering. Cleaning out formwork is required just as for pours carried out in the dry. Underwater, a vacuum tube operated by compressed air is usually used by divers for cleaning out formwork and it is extremely effective in removing any debris.

Placing concrete underwater is a specialist task and should only be carried out by fully experienced personnel. The conditions of working will normally be very difficult, as the divers work sometimes by touch only, and lighting is ineffective in cloudy water. Tidal and wave forces have to be contended with. Guidance on this can be found in BS 6349.[49] Other useful information can be gained from the latest CS technical report on underwater concreting.[50] Figs 76 and 77 show examples of concreting underwater.

5.3. Flexible forms and mattresses

Pumping is essential when filling the flexible forms and mattresses which are increasingly used for sea defence works, structural repairs and other underwater applications. The cost of temporary works for underwater repairs can be extremely high and may exceed that of the actual repair. For example, repairs which have to be carried out in the dry would entail a cofferdam of steel sheet piling and the structure may require bracing with baulk timber or structural steelwork. It will have to be pumped out and kept dry and the steel sheet piling will eventually have to be extracted. All this is extremely costly. The additional time involved in the temporary works may be a critical factor, if, say, a harbour entrance requires repair work. Flexible forms may provide a cost-effective alternative. Even though the technology is relatively new the cost advantages are clearly available.

Flexible forms provide an alternative to traditional timber

Table 22. Test results from cores taken from pile jackets formed using flexible formwork (courtesy Geotextiles Ltd)

Identification	Water/cement ratio*	Cement: kg/m^3	PFA: kg/m^3	Compressive strength: kg/cm^2	Abrasion: cm^3/cm^2
Job test cylinders	0·45	635	—	395	0·018
Job test cylinders	0·41	602	237	467	0·116
Cores:					
test at surface	0·50	578	220	401	0·121
12·5 mm below surface	0·50	578	220	401	0·113
University of Chile	0·50	578	220	—	0·100

*As mixed and before expulsion of water through fabric.

Fig. 78. Flexible formwork support to large pipe ready to be filled with grout (courtesy Geotextiles Ltd)

and steel formwork. They provide a permeable container to support the concrete in its fluid state during hydration. Fixing is relatively easy. Flexible forms can be tailor made to suit the many different underwater repair jobs. The material used is a permeable woven fabric which controls the bleeding of water from the mix, while still retaining the solids. The manufacturers claim that this produces early stiffening and improves the density and strength such that compressive strengths of the concrete placed underwater by this method are higher than those placed in the dry. They also claim that there is improved concrete durability and resistance to abrasion.

In 1966, deteriorated concrete piles supporting a bridge at Abescon, NJ, USA, were structurally repaired with 30 in. (762 mm) dia. flexible formwork jackets. In March 1971 two 4 in. (102 mm) dia. cores were taken from these pile jackets which had been exposed to the severe weather which is often experienced in the area. These cores were tested for resistance to abrasion (ASTM C-418-64) and for compressive strength (ASTM C-39), one by a commercial laboratory and the other by the University of Chile. Table 22 shows data from these tests. The excellent abrasion resistance of the mortar was attributed to the fact that the expulsion of excess mixing water through the permeable fabric resulted in a substantial reduction of the water/cement ratio of the hardened mortar.

The fabric is available with various strengths and permeabilities. Low volume, simple shapes can be used without

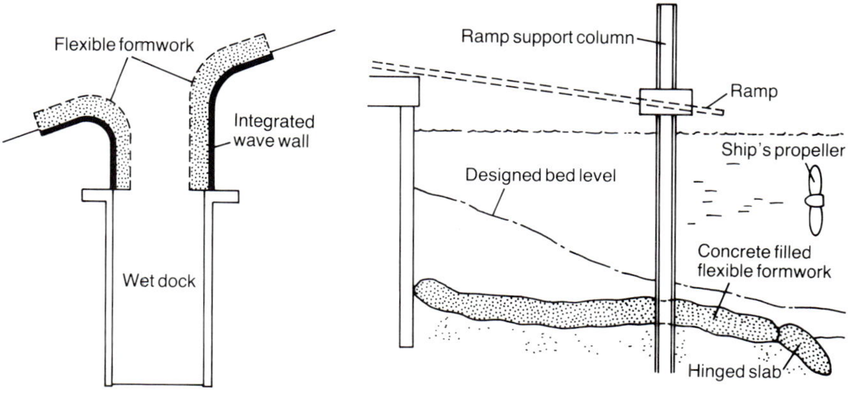

Fig. 79. Examples of flexible formwork repairs (courtesy Geotextiles Ltd)

external support, but larger or complex shapes, particularly those with high vertical faces, require an external framework to prevent distortion.

Flexible forms can of course be used above ground and their use is not mainly underwater. The material is extremely light and easily handled. It is left in place after the concrete has been placed, is not reusable and will offer no long-term protection to the concrete. An example of the application of flexible formwork is shown in Fig. 78. It should be noted that the concrete is pumped in at the bottom. Other typical applications are shown in Fig. 79 for docks and harbour work.

Concrete for flexible formwork

Pumpable concrete is suitable for filling flexible formwork underwater. The provision of the concrete should therefore be no problem (see section 5.2). The filling materials used in flexible formwork are not limited to concrete and will vary according to the application. All can be successfully pumped. Grout is used mainly for thin sections and for some types of structural repairs. Concrete is normally used for thicker slabs and shapes and may contain plasticizers. PFA cement replacement has also been found to be suitable.

Pumping the concrete is usually straightforward, but filling complicated shapes may require multiple injection points positioned at the bottom of the formwork and small delivery lines. In this case the concrete is forced upwards by pump pressure.

Mattresses

The original quilt type mattress was developed and patented in Holland in 1964. Grout filled mattresses of various styles are now used in many countries and have particular applications in the formation of revetments for sea defence works, canals and rivers. Mattresses can be installed easily and quickly and are exceptionally cost-effective. Since their introduction over 20 years ago improvements have taken place in placing techniques, in the textile material itself and in the mix design, producing an effective answer to the continuing problem of coastal erosion or scour of one form or another.

A double skin fabric is woven with the two layers connected at regular spacings, so that when grout or concrete is pumped between the two layers a continuous mattress is cast. The types most commonly used include woven filter areas located at the interwoven zones which afford relief of hydrostatic pressure in the embankment. Dual layer mattresses are available in thicknesses of 50–1000 mm. Tubular type mattresses (Fig. 80) are commonly used.

Mattresses are often laid on reclaimed land formed of dredged materials. High production figures for the laying of

Fig. 80. Typical tubular type mattresses (courtesy Geotextiles Ltd)

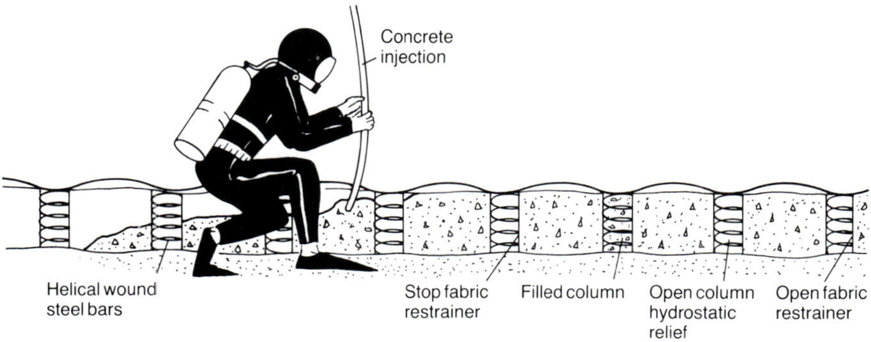

Fig. 81. Principle of the Ovolo matttress (courtesy Geotextiles Ltd)

such mattresses are claimed by contractors: an average of about 1750 m^2 per shift has been quoted.

Other types of mattresses are permeable in the finished state. The upper and lower fabric layers are connected by threads so that after filling, the total area covered by the mattress is occupied by grout or concrete. A further type allows movement to take place. The filter zones are arranged in straight lines, producing planes of weakness which in turn allow rotation to occur. Nylon cord reinforcement ensures that the mattress integrity is maintained after cracking has occurred, say from ground settlement. This facility of allowing movement is very important where the application is concerned with fill material, for example on a made-up embankment.

Most applications use sand/cement grout placed by concrete pump. As with flexible formwork, the permeable fabric allows excess mixing water to be released rapidly which means that the concrete sets quicker, particularly on inclined surfaces. Again, improved long-term strength and durability are claimed by the manufacturers.

The Ovolo mattress produced by Geotextile Projects Ltd, of Letchworth, Herts, has been widely used in underwater construction and repairs. This type of mattress repair was chosen at Larne harbour in Northern Ireland for propellor scour protection at roll-on, roll-off ferry berths where the scour problem was particularly severe. This type of mattress is modular and incorporates several special purpose features such as reinforcement and hydrostatic pressure relief facilities. Furthermore, the manufacturing method allows combinations of different fabrics to be used to utilize special physical properties. For example, if a mattress is required to fit snugly to a surface, which could be a dock wall or steel

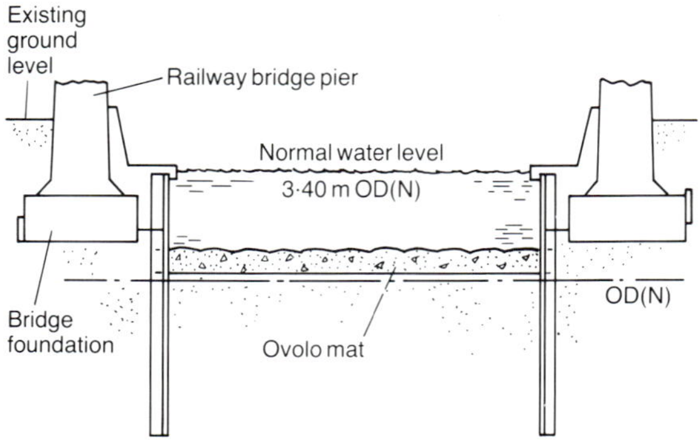

Fig. 82. Use of Ovolo mattress to form 0·5 m concrete slab (courtesy Geotextiles Ltd)

sheet piling, a fabric which stretches by up to 20% well below its ultimate tensile strength could be used. The mattress can be made to various thicknesses and used in several layers. The principle of the Ovolo mattress is shown in Fig. 81.

Figure 82 shows the work carried out at Ipswich, UK, in connection with the widening of a river below a railway bridge. The specification called for a 0·5 m thick concrete slab to act as a strut to support steel sheet piling at each side, to prevent any movement when widening took place. Full contact between the concrete and the pan of the sheet piles was essential. The concrete strength specified was 21 N/mm^2

Fig. 83. Mattress protection for sea defence work (courtesy Geotextiles Ltd)

at seven days. The Ovolo mattress arrangement met all these requirements satisfactorily and proved to be a very economic solution. The alternative in this case would have involved constructing a cofferdam and rerouting the water course.

It is clear that repair work and preventive measures in docks and harbours will continue to be needed because of the erosion problem caused by the wash of propellors, and the increasing size of the leisure boat industry, which is using larger and faster vessels. Mattresses and flexible forms are a cost-effective solution. It is specialist work which should be carried out only by experienced pump operators, who, in conjunction with the fabric manufacturers, will design and supply mattresses and flexible formwork to suit most clients' requirements (Figs. 83 and 84).

Fig. 84. Pumping grout into a mattress (courtesy Rapid Concrete Placing Ltd)

In recent years there have been considerable advances in geotextile technology which is applicable to the construction industry. This topic was given considerable coverage in the second international conference on geotextiles at Las Vegas, in 1982.[51] Lupton[52] describes in some detail the theory and applications which are applicable to flexible formwork.

5.4. Continuous flight auger piling

Concrete pumping plays an important part in the construction of cast-in-situ piles using the continuous flight auger system, commonly referred to as CFA piling, which requires the use of a medium sized trailer pump (Fig. 85). This type of pile was developed in the USA using grout as the pile material. Later the Japanese developed the system further, creating a market for this system of piling. In Europe, the method was first used in Holland where ground conditions are well suited.

In recent years this system of constructing small diameter cast-in-situ piles up to 600 mm diameter has been increasingly used in many parts of Europe as well as the UK and is popular because it offers advantages in economy, reduced noise levels and vibration as well as speed of placing. The method involves the use of a hollow stemmed continuous flight auger which is fitted with an expendable shoe or cap at the base. The auger is drilled into the ground to the required depth which has been predetermined by soil engineers. Concrete is then pumped through the hollow stem, and the concrete pressure forces the cap off. The auger is

Fig. 85. CFA piling rig (courtesy GKN Keller Foundations)

slowly withdrawn at a controlled rate while concrete pumping continues until the base of the auger reaches ground level. The small amount of soil displaced to ground level during boring is quickly removed to reveal clean concrete at the head of the pile before inserting the reinforcing cage (usually with the assistance of a vibrator) into the still fluid concrete.

CFA piling appears deceptively simple, yet its success demands good assessment of ground conditions, control of the boring process and (most importantly) control of the concrete delivery during withdrawal of the auger. The integrity of the pile shaft is dependent on concrete being supplied to the base of the auger stem at the rate required to form a full pile section as the soil laden auger is withdrawn from the ground. Problems have been encountered in the past where this balance has not been achieved and this was noted in the technical press.[53] However, in recent years such

Fig. 86. Instrumentation for the CFA rig operator (courtesy GKN Keller Foundations)

problems have been overcome and the acceptability of the system has been enhanced by the provision of sophisticated instrumentation to assist the pile operator in the control of pile construction.

By understanding the risks associated with CFA piling (and perhaps to a certain extent from past experience), several of the major foundation engineering companies in the UK have developed computerized monitoring systems which are able to provide the rig operator with on-the-spot visual information, to enable the whole operation to be controlled properly. The information is available to the resident engineer or clerk of works and is thus an invaluable aid to supervisory control of the work.

In addition to the traditional pressure gauge in the concrete supply line, special sensors can monitor depth, rate of boring, concrete flow, extraction rate of the auger and volume of concrete pumped, all against time. GKN Keller

Foundations is one such foundation engineering company, based in the UK, which has developed a computerized monitoring system.[54] The instrumentation installed in the cab of GKN rigs consists of a microcomputer with a keyboard/display unit and printer (Figs 86 and 87). The movement and position of the auger head during drilling, and the rate of flow, volume and pressure of the concrete are recorded accurately. A print-out showing a record of each pile is produced while the pile is being constructed; individual pile records are available within seconds of finishing each pumping operation. Recorded information can later be fed into the main office computer for future reference. It is of course important that the operator is properly trained and that the design engineers have full information on the soil conditions existing at every site — a prerequisite of any foundation project.

Other ground engineering companies have developed and are using monitoring systems. For example, the French firm Jean Lutz SA has produced an instrumentation package for recording measurements during the drilling and concret-

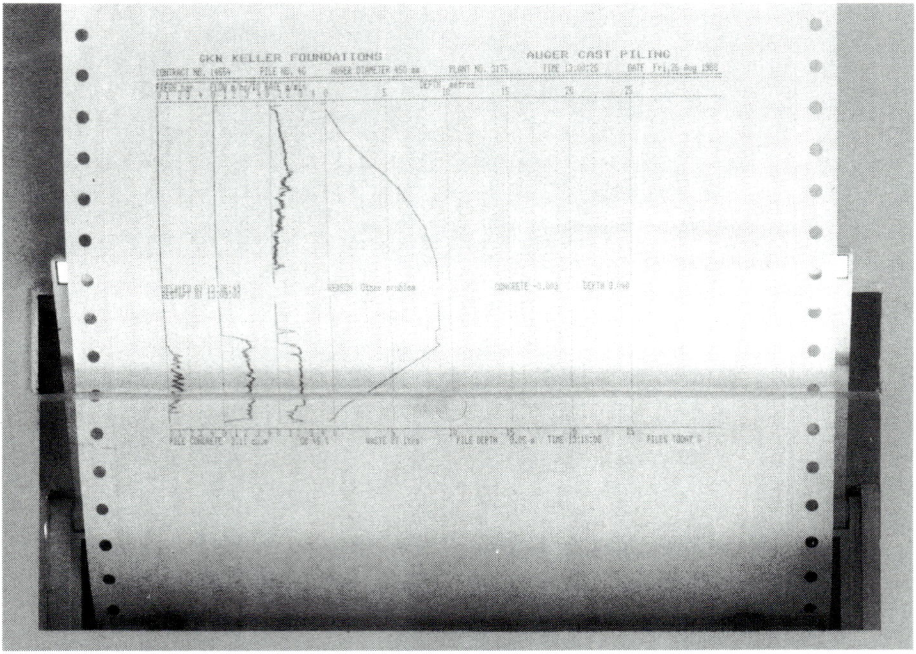

Fig. 87. Print-out immediately available, giving a record of each pile (courtesy GKN Keller Foundations)

Fig. 88. Typical trailer pump used for CFA work (courtesy Premier Concrete Pumping Ltd)

ing phases of the pile construction. This system can produce an interpretation of the pile section as constructed, when the data are fed into an analytical computer package.

The type of concrete pump used in CFA piling is normally the medium sized trailer pump with an output between 20 and 40 m^3/h (Fig. 88). They can be hired readily although some piling companies own pumps. These pumps are easy to handle and manoeuvre on site, very reliable and relatively simple to operate. Pumping pressures required are well within the ratings of such pumps and the pressure at the underside of the auger head is normally under 5 bar.

Close liaison between the rig operator and the pump operator is essential during construction as the rig operator can see from the instrumentation display panel in the cab what is happening and must be able to relay signals to the the pump operator as and when required. The rig operator will normally do this through the driller/ganger who is near the pile position in view of both the rig and pump operator. It is obvious that a clear system of signalling must be agreed before the work starts.

The other usual safety precautions taken during concrete pumping (see chapter 12) are equally applicable to CFA piling. In addition, as there is generally more flexible hose used in CFA piling than in normal concrete pumping, special care has to be taken to avoid the hose kinking and handling hoses in general. Hanging hoses should be fitted with safety chains or strops which will prevent them falling and causing a serious accident in the event of a coupling failure.

Pump operator training on the entire system is essential, as the operator should be aware of what is happening at all times during the construction of the pile and, in particular, that this is not a simple concrete placing operation with an open ended delivery hose.

Concrete mix design for CFA piling will follow the normal criteria for pumpable concrete. Aggregate sizes of up to 20 mm can be used. The cement content is usually not less than 340 kg/m^3 and a plasticizer may be used to increase workability.

Integrity testing and/or load testing of piles is normally carried out on most sites and pile failures are now rare occurrences. It is particularly interesting to analyse and compare integrity test results with the monitoring results. Pile cross-section variations shown on print-outs may be correlated with the sonic signal reflections from integrity testing.

The provision of sensitive monitoring has offered scope for improved techniques and has largely removed any lingering doubts about CFA piling, but information still needs to be exchanged freely between contractors and engineers. The costs involved for the firms who have been prepared to provide sensitive monitoring systems has been considerable. The use of such monitoring systems is an additional cost and in fairness the client should be prepared to accept this. The ICE *Specification for Piling*[55] calls for contractors to submit details of their monitoring methods, but eventually it is surely sensible that specifications should demand recognized monitoring systems.

The role of the concrete pump in CFA piling is extremely important as the placing of concrete under pressure in a controlled manner is an integral part of the whole operation. As the use of this system of piling increases there may be scope for the development of a unit specifically for this purpose.

6

Cost comparisons

Comparing the costs of placing options is something which is of extreme importance in the construction industry. It is equally important that all the factors involved are taken into consideration including the effects on the contract as a whole. Only then will a true cost comparison be made.

6.1. True cost comparisons

If a project manager in the UK is asked the preferred concrete placing method, some would reply that their preference would be for the concrete pump which was always given first consideration, adding that cost comparisons were made before a final decision. Many will say that if the pour is outside the range of the crane, if the pour is above a certain size or if the access is difficult, a pump will be used. Other typical replies are

(*a*) 'If we have a crane on site (anyway), why not use it? We're paying a weekly hire for it.'
(*b*) 'If we use a concrete pump, it's an additional cost.'
(*c*) 'Using a pump is quicker, but we still gain very little time on formwork striking, if any at all.'
(*d*) 'We work out our costs and the pour has got to be a certain size before a break-even point is reached and the pump becomes viable.'
(*e*) 'Pump hire rates are high enough for actual pumping time, without having to pay the same hourly rate if it is standing waiting for concrete.'
(*f*) 'We don't like the minimum hire period charges.'
(*g*) 'We object having to pay extra for a pumpable mix.'

These comments might be expected from the construction manager involved in the day to day running of a contract. But statements (*a*)–(*d*) can only be proved to be right or wrong if proper or true cost comparisons are carried out.

Statements (*e*) and (*f*) might well disappear if adequate planning is carried out with the contractor taking the lead. Statement (*g*) is more often than not a small enough increase to be insignificant and in many countries does not arise.

The speed at which concrete can be placed depends on a number of factors: location and type of pour, concrete supply, speed of compaction, finishing and site conditions. Therefore it can be misleading to base any cost comparisons of methods of handling concrete on theoretical or potential outputs of the concrete pump. The total cost of a concreting operation can only be estimated by taking into account all the factors for all the possible methods of handling concrete for that particular job and converting the figures into costs.

Illingworth[56] devotes a chapter to comparative costing, clearly illustrating the need for what he terms the three stage cost comparison, which should include the whole of the concreting operation i.e. mix, haul and place (Table 23). Illingworth summarizes the cost comparison technique into three main areas

(*a*) cost comparisons of alternatives on a like-for-like basis
(*b*) consideration of intangible costs
(*c*) influence of the method on other operations and the contract as a whole.

He includes case studies and detailed examples of how such comparisons can be quantified accurately.

Accurately comparing costs of various placing options on a direct like-for-like basis will involve what can be termed real labour and plant costs to the contractor. For example, the wages of the placing crew are not the actual cost as all

*Table 23. Three-stage basis for true cost comparison for whole concreting operation**

1. Mix	Site batch or ready-mixed concrete supply direct to crane (skip), dumper, pump
2. Haul	Crane, dumper, pump to point of placement
3. Place	Labour and ancillary placing equipment (vibrators, screeders)

* If ready-mixed concrete truck discharges directly into formwork, 1 and 2 are combined.

the additional costs such as insurance, paid holidays, subsistence, travelling, etc., should be included. This in fact can more than double the hourly rate. When plant rates are considered, all costs must be included. For example, for a tower crane it is essential to include costs of fuel (oil or electricity), repairs and maintenance, on and off site haulage charges, anchorages and/or track, foundations, erection and dismantling, and provision of electrical supply equipment and cables. The contractor has only to make a rough calculation of any of these charges to be reminded that they are indeed significant.

The intangible factors as described by Illingworth are those which are very relevant to a true cost comparison, but on which it is difficult to apportion cost figures. Their influence on the final figure can be decisive, particularly when the initial direct appraisal reveals little difference in cost. These factors would include

(*a*) the effect of delays on plant standing charges, including shuttering and scaffolding
(*b*) influence of a concrete pump on the selection of other plant
(*c*) problems of restricted sites
(*d*) possible overall time savings in the contract
(*e*) influence of damages clauses.

Contracts which are plant intensive can be seriously affected by delays. Careful consideration of this aspect should always be made when comparing costs of alternative placing options. Savings could be substantial for one site, yet relatively insignificant on another: the cost of delays must always be taken into consideration.

Illingworth[57] gave examples of the cost of stop-ends expressed as a percentage of the cost per cubic metre of concrete placed. The dramatic reduction in costs for high volume and large areas is shown in Table 24; the figures were taken from case studies. The specifying of bays based on the middle third rule will without doubt create a situation where stop-end costs will be a significant percentage of the total cost of the slab.

The decision to use a pump may influence the selection of other plant. For example, a tower crane may still be needed for certain operations, but now may be a smaller one and on a large contract one crane will suffice instead of two.

The problems arising from restricted sites which are so

Table 24. Examples of stop-end costs

Method of placing	Cost of stop-ends: %
Large concrete raft (2210 m^3)	
Poured in 50 m^3 bays	10·27
Poured in 3 bays	1·56
Large area 300 mm thick floor pours	
Poured in 40 m^3 bays (as specified)	9·38
Poured in 120 m^3 bays (as modified)	2·91

prevalent in urban areas can also be greatly eased by choosing a placing method which is quicker, e.g. the ready-mixed concrete trucks have a quick turnaround if a pump is used instead of the crane/skip method. Also the choice of using ready-mixed concrete and avoiding further restriction of an already congested site by taking up space with site batching plant, can be yet another influencing factor to be considered when calculating cost comparisons. Most contractors who have operated in restricted sites know how costs for all site operations can escalate from materials handling problems and of the frustration felt at times by the work force. It may be said that such costs should be included in the tender figure, but it is debatable if realistic sums are always included.

Thus the full implications of cost under the headings direct comparison of alternatives, intangible costs and influence on other work, must be assessed before it can be said that one option is cheaper than another. If this could be done at the tender stage it would be ideal, but the relatively short periods given between issue of tender documents and the return date make this extremely difficult in most cases.

Using a particular method of placing concrete may produce possible overall time savings in the contract. This might mean using a concrete pump and so release a crane for other work such as handling formwork and steelwork, both of which are expensive and labour intensive. A cost calculation may produce savings in overheads, etc., which are related to time, i.e. the length of contract may be shortened. In the end this saving might well more than cover any additional cost involved (Fig. 89).

Many contracts carry liquidated damages clauses and the damages for one week's delay could be several thousands of pounds. The influence of damages clauses on site management is given scant treatment, yet it is extremely important. The discipline imposed on management and supervision, when a concrete pump has been hired and concrete has been ordered for a specific time, should normally be very effective, because there is the threat of claims if the pump is standing idle or the ready-mixed concrete trucks are delayed. Consequently, one would expect more effort to be made to be ready on time. Unfortunately, at a time when competition is very keen and the pumps and ready-mixed concrete plants are underutilized, claims on contractors are rarities, yet understandable in the circumstances.

The effect of the option to use a concrete pump on other work is important in relation to formwork and steel-fixing. An increase in concrete production may mean that extra formwork is required with more labour involved to fix it. This may be for a shorter overall period and detailed calculations of time and quantities of materials are necessary. Steel-fixing quantities may be such that more labour is also required, again for shorter periods. In this event the steel-fixing subcontractor will normally welcome larger quantities which will enable the subcontractor to carry out the work more easily. Steel-fixing gangs can usually be recruited fairly easily in the UK.

By using a placing method, e.g. concrete pumping, which enables larger pours to be carried out, stop-end costs can be significantly reduced. It is not possible to give average figures for stop-end costs or to state the savings which can be made, as every pour is different. It has been proved in the past, however, that the savings in stop-end costs by casting in larger bays, which would be possible if the concrete was pumped, more than offset the additional cost of the pump in a basic comparison with an on-site crane.

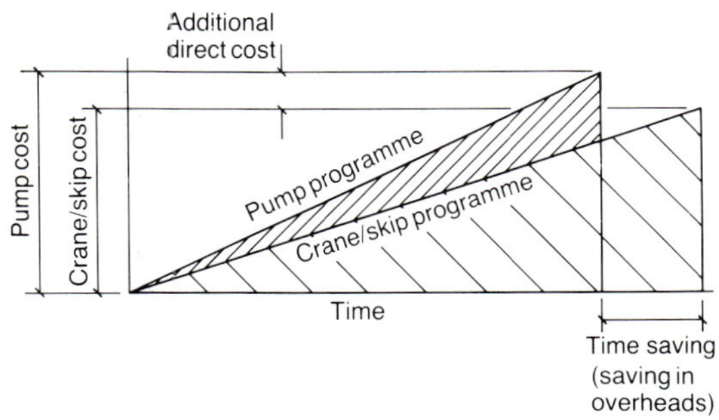

Fig. 89. Time saving: concrete pump against crane/skip

6.2. Cost comparisons: concrete pump and crane/skip

The Lancaster University concrete pumping report[1] produced figures which gave costs of labour and plant (excluding ancillary plant such as vibrators, and screeders) for placing concrete by both pump and crane. The costs calculated were direct (basic) costs only.

Averaging the UK and West German figures, the cost per cubic metre placed by the crane/skip option was cheaper; roughly 80% of the cost of the concrete pumping option. The average pour size for the latter method was almost double the size of the crane/skip sample, which was 25 m^3. Cost comparisons therefore between these two options will indicate that it is cheaper to use a crane if the calculations are limited to direct costs. The uninitiated might well decide to go ahead with the apparently cheaper method. It is at this point that the previously described associated cost factors must be considered and a true cost comparison made. The result may then be surprising.

The size of the pour has a significant influence on

(*a*) placing costs
(*b*) placing outputs.

Figure 90 shows a hyperbolic curve is always produced when plant such as a concrete pump or a crane is used for a single activity, e.g. placing concrete. In this case as the pour size decreases the unit cost increases. Below 40 m^3 the cost rises steeply, indicating that there may be a pour size limitation for this method. Break-even points as suggested by

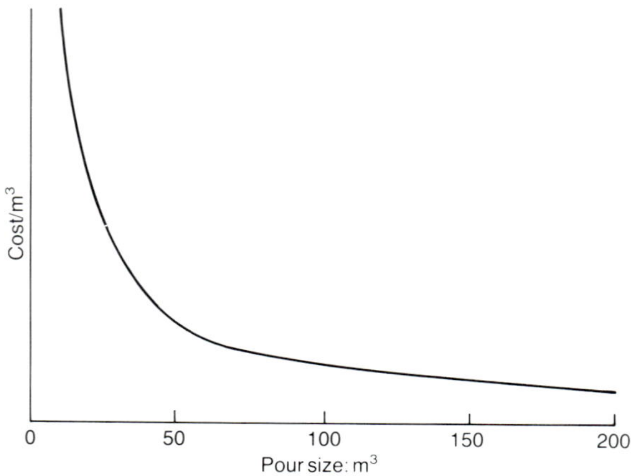

Fig. 90. Cost/pour size

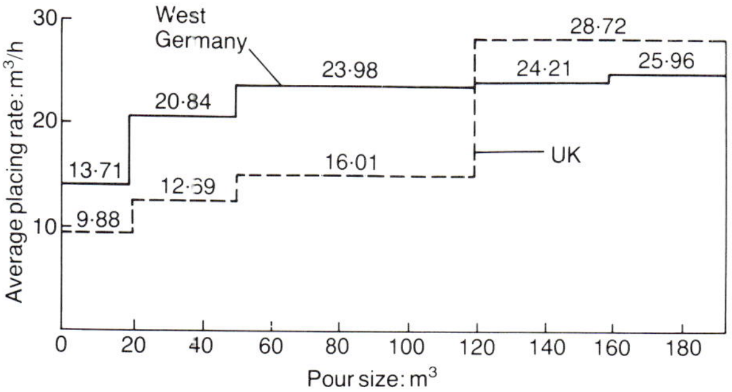

Fig. 91. Average placing speed (pumped pours) in the UK and West Germany

some project managers (see section 6.1), are probably calculated on the basis of such a graph, but the full implications are not being realized. For larger pours, unit costs decrease only marginally and there is a point where there is little advantage, because higher outputs means excessive wear on the pump, with corresponding additional costs off-setting the savings.

Average placing speeds for various pour sizes are illustrated in Fig. 91. Outputs in excess of those shown in Fig. 91 will have been achieved from time to time, but case studies in the Lancaster University study[1] produced average figures for ranges of pour sizes which were recorded in West Germany. The average production rate for crane/skip pours was just over 12 m^3/h, although the sizes of pours recorded by this placing method did not exceed 50 m^3. There is no doubt that when considering an average size pour under normal placing conditions, the placing speed which can be anticipated using a concrete pump will be at least twice as fast as using a crane.

It is apparent, therefore, when comparing the costs of a pump and a crane/skip, the cost differential in favour of the latter method using basic costs is cancelled out between pours of 50 and 100 m^3 as the time saving becomes more significant. Production output and pour sizes vary very little using the crane/skip option, but show a steady increase when a concrete pump is used. Fig. 90 does not quantify the time savings which can result from using a pump, but it does indicate the potential time saving. No decision on placing methods should be made before the time and hence cost savings are very carefully considered and evaluated.

The cost ratios of plant to labour vary with pour size. Table 25 is compiled from current average pump hire rates.

Table 25. Plant/labour cost ratio: UK compared with West Germany

Pour size: m^3	Speed of pour: m^3/h					
	30		20		10	
	UK	WG	UK	WG	UK	WG
250	1·44	1·30	1·32	0·87	1·19	0·60
200	1·67	1·51	1·38	0·96	1·22	0·61
100	2·01	1·62	1·70	1·08	1·38	0·62
60	2·70	1·70	2·10	1·10	1·60	0·65
20	7·90	2·59	5·30	1·73	2·60	0·86

The marked difference in ratios between the UK and West Germany at the lower range of pour sizes (below 60 m^3) is largely due to the minimum charges levied in the UK. As the pour size decreases, the ratio of plant cost to labour cost increases. In other words, the plant costs will rise more quickly than the labour costs as the pour size becomes smaller. This is what might be expected. A similar graphic configuration would appear if crane/skip figures are used.

All this indicates that pour size is a considerable influence on placing costs. It follows, therefore, that the client looking for a job which is as economical as possible and the contractor looking for minimal costs, should be aiming for larger pours. The reluctance of some engineers to permit design variations which facilitate large pours is regrettable, because in the end it will increase the overall cost of the project and the completion date will be extended.

With advances in concrete technology it is possible to design for large pours which, with the help of a concrete pump, can be completed in a single day. Hitherto this may have had to be staggered over several days to cope with curing and stop-end stripping time.

The main principles to be followed for successful continuous casting of large pours can be found in the C&CA current practice sheets produced in 1976 and 1977,[58–60] which deal exclusively with large pours. The important points to remember are

(*a*) control of internal thermal strain
(*b*) avoid external applied restraints
(*c*) cast the complete section in one pour.

In current practice sheet 28,[58] the use of concrete pumps is recommended to meet the requirements of continuous casting, which includes completion as quickly as possible. Concrete sections up to 35 m long, 4 m deep and 3000 m^3 in volume have been cast.

Size of placing crews

The number of people in a concrete placing crew, whatever method of placing is used, can vary between four and twelve (Fig. 92). The number required depends on several factors. It is also not possible to state categorically if more people are required for a concrete pumping pour than for a crane/skip pour, even if it was for the same pour. Nor is it possible to say that a specific number is required for various types of pour e.g. slabs, beams, columns. As a general rule, however, a thick slab requiring the same finish as a thin slab may be satisfactorily completed with say one person less, as less time per cubic metre is required for finishing. Beams and columns may place constraints on placing space and fewer people will be required, but for a longer period. Boxing out or upstands in a slab will make placing and screeding more difficult. All such items have to be considered.

Fig. 92. Typical placing crew for a pumped concrete pour (courtesy Steetley Construction Materials Ltd)

It is argued sometimes that placing by pump is quicker, therefore more people are required to keep pace with the supply of concrete. But distribution by pump is easier than when using a skip and consequently fewer people should be required. When a crane is used, the crane driver is an additional person, but a company owned pump will require a company employed operator.

To calculate the number of people required, the speed of placing, type of pour and finishings must be considered. Start off with a basic placing crew of four (one person for each operation of handling the hose end, spreading, vibrating, finishing), and add additional people to meet the specific requirements as indicated above. This exercise must be carried out when any cost comparisons are made. It is not difficult but is absolutely essential. It must also be remembered that the labour considered is to be associated with the use of dominant plant (see section 1.3), i.e. a crane or a concrete pump. One thing is certain — too few people will prove extremely costly!

6.3. Concrete pump operating costs

It is appropriate when considering costs in connection with concrete pumping to examine the main items of cost in the operation of the pump itself and to look at the effect of utilization on profits. The figures given in Table 26 are for a mobile pump which is hired out on an hourly basis. It is important to differentiate between fixed and variable costs. The percentages quoted are based on a recent UK survey. Although the figures may vary from company to company and will also depend on the type of pump, they are accurate enough to illustrate the main costs. The pump cost is based on a new pump depreciated over six years on a straight-line

Table 26. Operating costs of a mobile hire pump

Item	Percentage of total cost per year
1. Capital cost of pump including finance charges (if applicable)	33
2. Operator's wages (cost to the company)	27
3. Fuels and oils	20
4. Overheads, insurance, etc.	10
5. Maintenance and spares	10

calculation. These are the percentages of total cost required to break even. To make a profit, the turnover has to be increased.

The utilization on which these figures were calculated was based on an average of six hours working per day, i.e. six paid hours for the pump hire. An increase in utilization of 50%, i.e. up to a full working day of nine hours would have the effect of increasing items 3 and 5 as more fuel would be required for longer operating hours and more wear and tear on the pump would mean extra costs in spares. It is not strictly correct to say that the other items, particularly the operator's wages, would not be affected, but broadly speaking it can be assumed that any differences are relatively small and may be disregarded for the purpose of this exercise. Thus, items 1, 2 and 4 may therefore be taken as fixed costs.

An increase in work-load is going to increase items 3 and 5 which together form only 30% of the total cost. Therefore costs will increase an indeterminate fraction of 30% of the previous total costs. The turnover should theoretically increase by the same amount as the utilization increase, giving the pump hire company a substantial profit.

Putting this into the simplest form of algebraic expression, let the total cost of the pump per day $= y$, the turnover $= x$, and the profit $= z$, then

$$x - y = z$$

If a break-even situation exists, $x = y$ and $z = 0$. But if x is increased by 50% and y is increased by 50% of 30% = 15%,

$$z = 1{\cdot}5 - 1{\cdot}15 = 0{\cdot}35$$

These figures illustrate certain points

(*a*) The reason why the pump hire companies require a minimum number of hours hire: this is usually five hours.

(*b*) The contractors hiring the pump should make the pours as large as possible so that they are not penalized by minimum charges.

(*c*) Hirers should arrange for several pours the same day if the individual pours are small. This can very often be done by a little careful planning with minimal overtime for formworkers and steel-fixers.

(*d*) Logically a reduction of hourly hire charges should follow if more utilization of the pump is guaranteed.

Fig. 93. Support pylon for bridge over the Rhine at Flehe, Dusseldorf, where concrete was placed by pump (courtesy F. W. Schwing/Burlington Engineers)

With regard to (*d*), most pump hire companies would be willing to reduce their hourly rates provided that they were guaranteed sufficient work on any one day. Trading constantly on minimum charges is not welcomed or indeed profitable in the long term. Thus the contractors who hire pumps would benefit if they took steps to increase the size of pours or the number carried out on any one day. The same effect would result from a pump being able to carry out pours on different sites on the same day (which is possibly more attractive to the pump hire company), but this would require a vast change in attitudes and planning approach from contractors generally (see chapter 7.2).

Typical example of cost comparison

The site management at the new Rhine Bridge at Flehe, Düsseldorf, were faced with the option of using a crane/skip or a concrete pump on the construction of a massive support pylon which was 145 m high (Fig. 93). Cost comparisons were carefully carried out.

A Liebherr tower crane 90.1HC was considered. Using two 2500 l buckets in tandem would have produced a placing rate of only 10 m^3/h and this was discounted because of the relatively low output. Instead it was decided to use a concrete pump which was able to place the concrete at an average rate of more than double the crane/skip method. Placing speeds averaging 25 m^3/h in the middle of the structure and 20 m^3/h at the top were achieved.

The cost comparisons included not only the direct costs of crane and pump but also the time savings and associated costs on the job as a whole. Using the crane for long periods on concrete placing would have seriously disrupted other work. The relatively high volume of concrete in the many sectional pours, from 110 m^3 at the bottom to 25 m^3 at the top, ruled out the use of the crane with its maximum 10 m^3/h output.

6.4. Other placing options

So far reference has been made mainly to concrete pumping and crane/skip options, but there are, of course, many other methods of placing concrete and combinations of methods open to the contractor. Even today the wheelbarrow still has a place and in the USA there is a motorized version.

Conveyors and hoists, as well as aerial cableways, still

have their place in concrete placing. Hoists in particular can be very effective in certain instances, for example for placing concrete in tall structures of limited cross-sectional area, such as bridge columns. Here a hoist equipped with an automatic discharge and with an adjustable chute at the placing position can prove to be the quickest and cheapest placing method. Such high-speed automatic hoists with a ready-mixed concrete supply at ground level, i.e. the truckmixers can back up and discharge straight into the hoist, will achieve very high production outputs. The adjustable chute will be able to place the concrete exactly where it is required provided that the dimensions of the pour are not too great. Such an arrangement is shown in Fig. 94 and can be used to advantage with slipform shuttering for certain types of towers or bridge piers. Hoists are readily available and relatively easy to hire or purchase. In the right circumstances they can be most effective.

Combinations of options should also be considered. For example, using a hoist for part of the transportation which

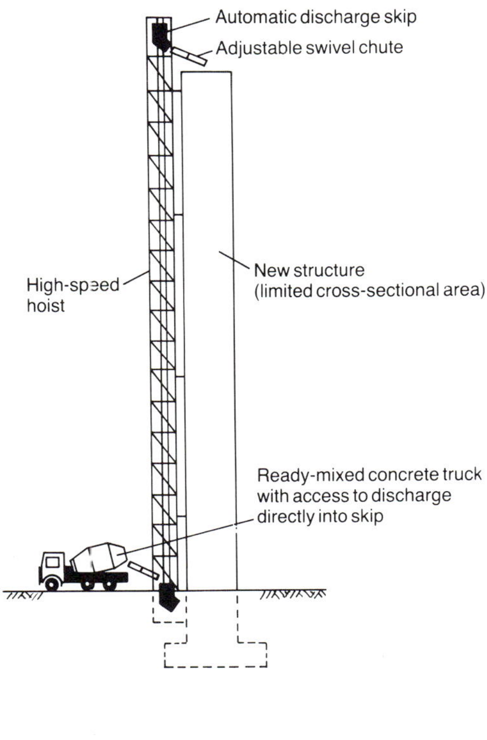

Fig. 94. Using a hoist for effective concrete placing operation

otherwise would have been carried out by say a tower crane, will obviate the need to use the cumbersome lifting process of the latter. The crane can be used to take over for final distribution to the placing positions. There is scope for ingenuity by considering rotating chute arrangements which, if set up correctly, will cover a surprisingly large area and so obviate the need for a crane. Use of separate placing booms or distributing arms in conjunction with concrete pumps are other combinations. Even direct placing by the ready-mixed concrete truck can be used more often and most effectively if there is proper access to the pour site for the truck. Placing costs are minimized in such cases.

The list of the ways and means which can be adopted for placing concrete is almost endless. No one can state categorically that one method or combination of methods is best for a type of pour, e.g. suspended slab, bridge columns, mass foundations. The method chosen will depend on a number of items specifically related to the pour in question, i.e. size, access and location. Provided that all these are taken into consideration, together with the effects the chosen method will have on the cost of the contract, the correct answer to the question 'Which is the best option?' will eventually appear.

7

Concrete pumping operations on site

Successful concrete pumping depends on having the correct pump and associated equipment for the job, together with adequate preparation and planning, a satisfactory supply of pumpable concrete and the necessary expertise in placing it. Each party to the pour has certain responsibilities.

7.1. Using the correct concrete pump and equipment

The potential buyer or hirer of a concrete pump will have no lack of advice as to the most suitable size and type from the manufacturer or hire company. It is important, however, that the supplier is given a clear indication of the full requirements which are to be met, i.e. lengths, heights to be pumped, typical mix specification, average placing rates. Such information is necessary for a proper decision to be made on pump capacity requirements. For example, large contracts in which a large amount of concrete has to be placed usually require high-pressure static pumps. The type of construction will also indicate the type of pump. On large power station contracts it has been the practice to use several static pumps strategically placed on site, which pump over varying lengths and heights. In this case it is normal to have fixed horizontal and vertical pipelines which are adjusted (either shortened or lengthened) for specific pours.

The pumps can be used in conjunction with separate placing booms positioned inside a partly constructed structure, or rotary placing arms as described in section 2.2. This ancillary equipment can greatly facilitate placing of the concrete. It is not unusual on such contracts also to use one or more mobile pumps for placing the smaller pours round the site which can readily be handled by this type of pump, thus avoiding the need to move the static pump or to dismantle and refix the pipeline.

Planning for concrete placing, particularly on the larger

contracts, is normally carried out at the tender stage and the appropriate costs included in the tender figure. The prudent contractor will have discussed the requirements with the pump supplier and will be satisfied as to the adequacy of the concrete pumping equipment to meet the particular needs in respect of access and placing speed. It is unlikely that the equipment will be ideal for all pours and situations on a particular site; e.g. a pump capable of pumping at 80 m^3/h may be required to operate at only 30 m^3/h in certain circumstances, but in general the contractor can obtain pumps which are capable of meeting the highest demands.

On other sites where pumps are hired to deal with specific pours, it is again wise to consult the hire company, but at the same time it is useful for the project manager to be able to asertain the size and type of pump which is required. It should be remembered that every concrete pour is different and always must be considered as such. A 50 m^3 pour on one part of a site may present an entirely different situation in another area of the same site and a similar pump may not necessarily be the correct one in both cases. For example, placing rates or output requirements for a thin slab involving more screeding per unit volume would be less than for a thick slab of the same volume. Longer lengths of pipeline and greater pumping heights will mean higher pumping pressures and the pump must have the required capacity. The mix must be considered: a high slump flowing mix requires a lower pumping pressure than a low slump mix. Intricate formwork boxing out or heavily reinforced slabs will cut down placing rates and the pump can be matched to this. It is seldom, however, that day to day pumped pours on a large site are always placed by the most suitable pump as it is impractical to have a full range of pumps available all the time. For such sites the contractor will have chosen pumps which will be generally most suitable to meet the highest demands of pumping pressure and output, but the contractor will also have other pumps more suited to lesser demands.

The major concrete pump manufacturers have devised a method of calculating the size of pump required for various conditions and specifications. They have produced a so-called concrete pressure performance diagram or nomograph (Fig. 96) which can be used to provide the project manager with a theoretical calculation of the size and type of pump required. It should be used as a guide only.

Loadwick[9] also describes the process of pump selection

and produces a similar diagram compiled in detail from first principles. The pump manufacturers' diagrams have been compiled from data obtained on actual pumping operations over many years and the theoretical answers provided for a given set of pour circumstances are claimed to be accurate to within ± 10%. It is very important, however, that all the relevant details of the pour in question are listed and taken into consideration. Applying the relevant pour information to the nomograph (see section 7.3) will provide a theoretical figure for the pumping pressure required and thus the choice of pump.

7.2. Factors to be considered when choosing a pump

Normally the initial requirement is for the pour to be completed in a certain time, or in other words a placing rate established. If the rate is 40 m^3/h it is incorrect to order a pump with an output of 40 m^3/h and expect to achieve it. It is at this point that the following factors relating to the specific pour should be considered, all of which should be taken into consideration when choosing the pump

(*a*) placing rate required (m^3/h)
(*b*) work factor of pump
(*c*) horizontal length of pipeline
(*d*) vertical length of pipeline
(*e*) number and radius of bends
(*f*) diameter of pipeline
(*g*) consistency (slump) of mix.

First, the rated output of the pump, say 40 m^3/h has to be downgraded by what can be decribed as a work factor.[14] The reason for this is that site conditions will always decree that the pump cannot perform for the whole of the pour in ideal conditions without stoppages of some kind or another, which are not necessarily due to the pump performance itself. Various delays can occur on site as described in chapter 8. Therefore the output has to be downrated by the work factor before taking anything else into consideration. A typical value of the work factor is approximately 75%, i.e. the pump is only working at its full rating of 40 m^3/h for 45 minutes in any hour. Thus the amended delivery rating will now be $40/0{\cdot}75 = 53\ m^3/h$.

As the pumping pressure will vary depending on the length, height and diameter of the pipeline as well as the number and radius of bends, these factors must be taken into consideration next. Frictional resistances in the horizontal pipeline mean that additional pumping pressure is required for a given output. This additional pressure is proportional to the length and will be a maximum at the pump, reducing to zero at the discharge point. When using vertical pipelines an allowance has to be made for static pressure. As concrete is approximately 2·5 times the weight of water, the increased pressure due to a vertical pipeline is about 0·25 bar/m.

Bends or tapers, which invariably form part of the pipeline, are points which cause the pressure to build up due to deformation of the concrete. It is usual to convert various sizes of bends into equivalent horizontal lengths for convenience of calculation. Schwing[14] recommend that a 30° bend with a radius of 1 m should be assessed as a 1 m length of horizontal pipeline. A 90° bend would be 3 m and a group of bends of radius 1 m totalling say 300° would be 10 m of horizontal pipe. For the smaller radius bends of, say, 250 mm in placing booms, distributors, etc., a 90° bend can be taken as equivalent to 1 m. A three-section placing boom will have nine 90° bends and it is usual to calculate the equivalent length as the boom length plus 10 m.

Pipeline diameter is important. For a given flow rate, the velocity in the pipe decreases as the pipe diameter increases, with the result that the pressure will be lower.[9] The choice of pipeline diameter is normally related to pump size and to the aggregate size. Pump manufacturers will normally specify the minimum size of pipeline diameter and wall thickness to be used with a specific model. As a guide, a pump with an output of 30 m^3/h and with not more than 200 m of pipeline might well use a 100 mm dia. pipeline, but for lengths in excess of 500 m a 150 mm dia. pipeline would normally be considered.

The pipeline diameter should always be less than the pump cylinder diameter. With regard to aggregate size, a general rule is that the pipe diameter should be between three and four times the size of the largest aggregate, e.g. if the maximum size of aggregate used is 50 mm, the pipeline diameter should be at least 150 mm. The most common size of pipeline diameter used is 125 mm.

Lastly, consistency of the mix will influence pumping pressure. The stiffer the concrete is the higher the pressure required. The lower the concrete workability is the higher the flow resistance is and consequently the higher the pressure is.[9] There are, of course, upper limits of slump when a very high water/cement ratio can lead to segregation and consequently a blockage. Conversely, if the mix is too dry it cannot be pumped. Therefore the slump must be known and must be within the limits for pumpable concrete, as described in section 4.3.

In the following worked example all the factors which have been described are taken into consideration. It should be remembered that the pipeline pressure is being dealt with and this should not be confused with the pressure of the hydraulic system.

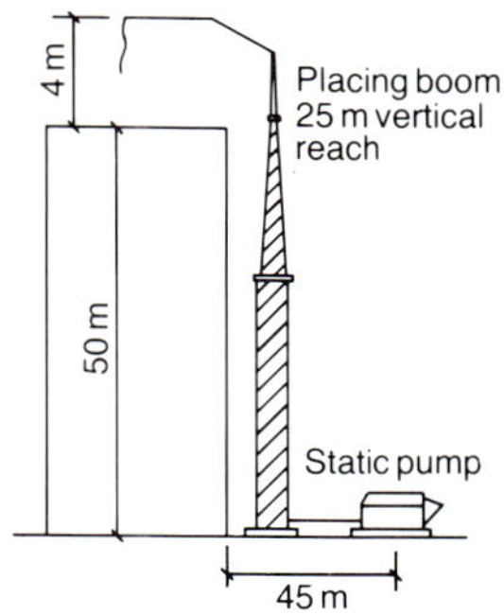

Fig. 95. Worked example

7.3. Worked example: choosing the correct pump

An example of a pumped pour is shown in Fig. 95. The details of the pour and requirements are as follows. The average output required is 30 m^3/h and the diameter of pipeline is 125 mm. The length of the pipeline, including the building, is 45 m horizontal and 50 m vertical = 95 m. The placing boom projects an extra 4 m vertically, and has a vertical reach of 25 m with four 90° bends and three 30° bends. The concrete slump is 75 mm.

Pump output

The work factor is taken as 0·75, and therefore the output required from the pump is 30/0·75 = 40 m^3/h.

Theoretical length of pipeline

The total length of the pipeline is 95 m. There are four 90° bends and three 30° bends making a total of (4 × 90) + (3 × 30) = 360° + 90° = 450°. Assuming that the bends have a radius of 1 m, 30° is equivalent to 1 m, and therefore 450° is equivalent to 15 m. The vertical reach of the placing boom is 25 m, and the bends in the placing boom are assumed to be equivalent to 10 m. Therefore the total theoretical length of the pipeline is 95 + 15 + 25 + 10 = 145 m.

The height the concrete has to be pumped to is the height of the building (50 m) plus the vertical boom projection (4 m), which is 54 m. The static pressure due to vertical pumping is therefore 54 × 0·25 = 14 bar.

Using the figures already given for output (corrected by work factor), pipeline diameter, theoretical pipeline length and slump, it is possible to arrive at a line pressure figure by using a nomograph (Fig. 96) to which should be added the static pressure.

Using the Nomograph

On the nomograph in Fig 96, locate 40 m^3/h on the upper *y*-axis and move across to the right to the pipeline diameter (125 mm), then downwards to meet the theoretical length of pipeline which was calculated to be 145 m. Now move horizontally to the left to intersect the slump line of 75 mm, interpolating between the 70 mm and 80 mm lines. At this point move vertically to meet the pumping pressure line. The reading at this point as shown on the nomograph is approximately 35 bar. To this should be added the previously calculated static pressure of 14 bar, giving a total of 49 bar. The pump chosen, therefore, should have a rated maximum concrete pressure of a figure in excess of this. Manufacturers will provide their recommended percentage to add, which is normally between 20 and 30%. In this example the pump required would therefore have a line pressure capability of between 60 and 70 bar.

7.4. Preparation and planning for the pour

Much has been said and written about the necessity of preparing and planning a concrete pour thoroughly if the pour is to be successful and profitable. Yet the fact remains

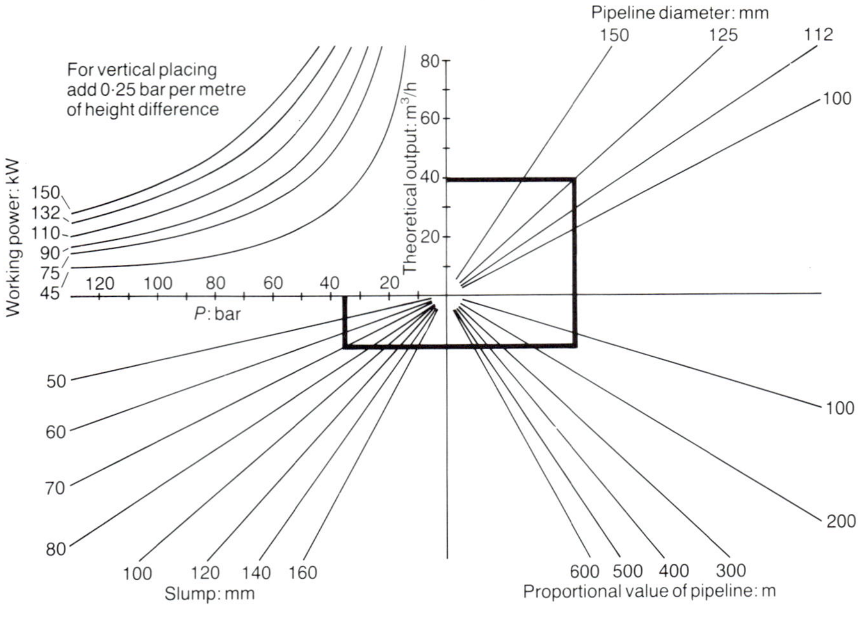

Fig. 96. Nomograph (courtesy F. W. Schwing/Burlington Engineers)

that serious delays do occur on such pours despite protestations from the parties involved that their part of the work, at least, has been thoroughly planned. The contractor who is sensible enough to draw up a written method statement of the pour requirements, which leaves no doubt as to the responsibilities of all parties, will have made the correct initial step towards ensuring the success of the pour.

There could be several firms involved with a concrete pour by pump. The number depends on whether the pump is hired and whether the concrete is to be supplied by a ready-mixed concrete company. In the event of the contractor using his own pump and his own batching plant on site (as sometimes occurs on very large contracts), the preparation and planning will still be needed. Only one organization is involved in this case, but co-ordination between the departments concerned is equally important.

As an example, a pour will be assumed where the contractor has hired a mobile pump and has the concrete supplied from a ready-mixed plant some distance from the site. Three firms are therefore involved.

Proper preparation and planning cannot take place without the necessary consultation between the parties concerned: the contractor, ready-mixed concrete supplier and the pump hire company. For this consultation to be successful it must be thorough and this will only be achieved if first of all a division of the responsibilities of each party to the pour is recognized and accepted. The responsibilities of these three main parties are varied but readily definable.

The contractor has the overall control of the pour and the role of co-ordinator. The contractor also has more responsibilities than the other two parties and the success or otherwise of the pour will rest largely on how well these responsibilities are discharged. The contractor should

(*a*) advise the other parties fully on the size and nature of the pour, particularly of the height, length and route of the pipeline
(*b*) give details of the specified mix stating that the concrete is to be pumped
(*c*) give the starting time, placing rate required and the estimated finishing time
(*d*) ensure that the pour preparation is fully completed by the start time and that it has been passed by the resident engineer or clerk of works

(*e*) provide an access road which is adequately prepared to safely carry ready-mixed concrete trucks
(*f*) provide a hardstanding at the pump position and also turning/parking places for truckmixers if required
(*g*) keep the access road clear for ready-mixed concrete trucks by arranging deliveries of other materials accordingly
(*h*) provide a wash-out area for the pump and ready-mixed concrete trucks
(*g*) provide bagged cement to make up the lubricating grout
(*j*) provide sufficient experienced labour for placing, particularly for handling the end hose
(*k*) make arrangements for replacing the crew during tea or meal breaks to ensure continuity of pumping
(*l*) have a spare vibrator, screeder or other ancillary plant available in case of breakdown
(*m*) arrange for adequate signalling methods if the pump operator cannot see the hose end
(*n*) make arrangements for regular slump testing and cube making as may be required, so that there are no hold-ups due to this
(*o*) keep the ready-mixed concrete supplier informed of any delays and telephone details of the final load quantity before the penultimate load is discharged
(*p*) assist the pump operator to lay out the pipeline, etc., and provide a banksman for the pump when required
(*q*) on a large site, give clear directions to the concrete supplier as to the correct entrance.

The ready-mixed concrete supplier is responsible for

(*a*) providing pumpable ready-mixed concrete as specified in all respects
(*b*) telling the contractor if the specified mix is not pumpable and agreeing mix design amendments
(*c*) arranging for the first truckmixer to be on site at least five minutes before the official start time
(*d*) maintaining the rate of delivery required by the contractor, by taking into consideration the distance from the plant to the site and the number of trucks required for this
(*e*) covering for breakdown of plant or trucks by making arrangements for alternative supply or replacement trucks

(*f*) regular slump testing and taking cubes.

The pump hire company is responsible for

(*a*) providing a reliable pump in good working order which is suitable in all respects for the type of pour
(*b*) the ancillary equipment: separate placing booms, distribution arms, pipelines, etc., which should also be in good working order
(*c*) providing a fully experienced pump operator who has been thoroughly briefed on the pour
(*d*) arriving on site sufficiently in advance of the official start time to allow enough time for setting up the pump and laying any necessary pipeline
(*e*) arranging for an alternative pump to be made available at short notice in the event of a breakdown
(*f*) ensuring a high standard of safety in the operation of the hire company's plant as well as monitoring the performance of the placing crews, particularly the handling of the hose end.

Interaction between the parties to a pour

Each party must be aware that it is dependent on the performance of the others and bear in mind that the more it co-operates the more chance there is that the pour will be completed successfully. A default by one will affect all three. Once it is accepted that co-operation is not only desirable but essential, the planning is made much easier and success is much more likely. The tendency is for claims to be raised against the guilty party in an effort to recoup losses. This is understandable, but is it the best way to solve the problem? The contractor who is delayed by the ready-mixed concrete supplier and says, 'It doesn't matter, we'll get our money back by claiming — we'll deduct our waiting costs off the invoice', is surely taking the wrong attitude. It would have been far better to have avoided the claim in the first place. How this could be done will now be examined.

The contractor, for example, may require the pour to be carried out on a certain day at a certain time. This, of course, is the contractor's prerogative. If the contractor instructs the ready-mixed concrete supplier and the pump hire company accordingly without discussion on the availability of truck-mixers, pumps, etc., for that particular day, the chances of obtaining a trouble-free pour are slim. The contractor should remember that these two firms are not merely materials suppliers, they are in essence subcontractors providing a service. They have daily schedules to meet and other customers with certain demands. They may feel obliged to accept the contractor's order and not turn away the business, but the timing of this particular pour may be far from suitable to them and may be an embarrassment. Ready-mixed concrete plants have a fixed number of trucks available and although they can supplement these from other depots, it is not always possible to implement customers' demands fully. Similarly pump hire companies have periods when there is a demand for a specific pump size and the company may have problems in fulfilling orders as it would wish. But in all fairness firms should make the contractor aware of any such difficulties.

If the contractor is prepared to be flexible and discuss pours well in advance of the date required, the contractor is far more likely to obtain better service and a claim situation will not arise. Many contractors favour an early morning start, say 07.30, often on a specific day. Friday is particularly popular because the concrete curing can take place over the weekend. The West Germans have a saying that *Freitag ist Betontag* (Friday is concrete day) and Friday mornings also appear to be very popular for pouring concrete in the UK. It might well be, however, that the contract programme would not seriously suffer if the pour was moved from an early morning start to say 13.00 or indeed to another day. It could be beneficial to the contractor, concrete supplier and pump hire company to fix the pour for a date and time which is more suitable for the concrete supplier and pump hire company, yet the date and time might still be acceptable to the contractor.

Such an arrangement is not easy, but compromises on start dates and times are possible provided that there is a willingness to be flexible and to co-operate with one another. This interaction between the three parties is fundamental to success.

Co-operation between the concrete supplier and the pump hire company is also of considerable importance. In some instances in the UK the pump hire company will not be aware of who the ready-mixed concrete supplier is until the first truckmixer arrives on site! Other countries (e.g. West Germany, where there is closer liaison between these two parties; sometimes the concrete supplier owns the pump) have a distinct advantage as co-operation is so much easier

Fig. 97. Co-operation between parties (courtesy Steetley Construction Materials Ltd)

and the result is that the contractor is more likely to get a better joint service (Fig. 97).

Close links between these two parties can only be a good thing, not only for any one pour on the day, but for other pours which may be scheduled for the same day and not necessarily on the same site. Problems might still arise, of course, but the chances of overcoming these quickly and effectively are greatly enhanced.

The Lancaster University recent research study[1] into concrete pumping revealed that the larger pours in the UK, over 100 m^3, invariably received detailed planning well in advance. Delay time on site expressed as a percentage of the total pour time was considerably less than for the smaller pours. It was also pointed out that there was a tendency to give scant planning to pours under 50 m^3. The fact that a small pour may only take say two hours, but has been delayed for any reason for one hour, means that the cost of labour and plant has increased by 50%. Overheads have also possibly increased. It is also possible that the materials cost will rise if premiums for truckmixers standing idle are chargeable under the agreement to supply. It should be borne in mind that the average size of concrete pour in the UK comes into this small pour category and the cumulative effect of delays to overall contract times and costs is significant. Illingworth[61] draws attention to the intangible costs in direct costing, a point which is particularly relevant in this context. It is discussed more fully in section 6.1.

It cannot be disputed that much of the delays experienced on pumped concrete pours could be significantly reduced if sufficient attention is given to more co-operation between participating parties.

7.5. The pour

The requirements of consultation and co-operation in planning for a pumped pour of concrete have been outlined in section 7.2 and the need for good communication on site can be taken as a logical continuation to this consultation and co-operation. Modern communications systems make it relatively easy to keep in touch and to keep people informed of what is going on. In the case of the concrete pour this means that all the people involved should be aware of what is happening and when. Unfortunately this does not always happen and even minor delays can become major ones. Before the pour starts the people involved at the site, the ready-mixed concrete plant and pump hire depot should have the relevant telephone numbers available and preferably also the names of the persons to contact.

It is a fact that valuable time can be lost trying to find a telephone number or locating the correct contact in an emergency. The site should be notified immediately of any problems at the ready-mixed concrete plant which could mean a delay in deliveries. It is equally important that the concrete supplier is informed of any delay at the site so that truckmixers can be diverted to other sites if possible. If the concrete supplier is not informed, there could be a queue of trucks waiting to be discharged, which can be a costly affair. The pump hire company's representative must also keep in touch with the company's depot to advise the company of any delays which might affect the timing of the next pour. A pump breakdown, of course, produces a real crisis and a repair or replacement cannot be effected quickly enough. Fortunately such breakdowns are rare occurrences.

A good start is important. A good start means starting at the programmed start time. It applies to all three parties to a pumped concrete pour. Delays at the start are common-

Fig. 98. Four mobile pumps placing 800 m^3 of concrete (courtesy Pochin Group plc)

place and are due mainly to the pour not being ready and to the concrete supply arriving late. It is very seldom that the pump is not ready at the appointed start time.

As the contractor has ordered a specific start time, the contractor must take responsibility if the pour is not ready. This is usually because approval has not been given early enough by the resident engineer or inspector. Some contractors leave it too late before asking for an inspection. On the other hand, the contractor may have originally ordered a start at 07.30, but on the day before the pour will telephone the ready-mixed concrete supplier and say, 'provisionally 07.30 start, but we'll telephone when we're ready'. This probably means that the start will be about 08.00: a half hours' delay for all concerned. Irrespective of the cause of the delay, the ready-mixed concrete supplier in both cases has to start rescheduling trucks for other pours and the pump hire company will have started to worry about possibly being delayed for the next pour that day. Other sites which are going to be supplied from the same ready-mixed concrete supplier become involved when it is necessary to reschedule truckmixers and these sites are most likely to feel the adverse effects of delays in the supply of their own order. These sites will be delayed through circumstances outside their control. This is described in the *Manual of ready mixed concrete*[47] and

Fig. 99. Small pumped concrete pour at Sizewell power station (courtesy F. W. Schwing/ Burlington Engineers and John Laing Construction)

is called the cascade effect, which can be quite dramatic by the end of a busy day.

The concrete may arrive late simply because the ready-mixed concrete supplier has taken on too many jobs which are starting at the same time and has an insufficient number of trucks to cope with this peak period. It frequently happens that one or two truckmixers will arrive on time, but then there is a lengthy delay, which is a sure sign that other jobs are starting simultaneously.

Reflecting on these common delays at the start of a pour, it becomes obvious that these could be largely eliminated if the parties concerned were prepared to co-operate more fully and show an awareness that should they create a problem or delay then others are affected as well. Every effort should be made, therefore, to make a start at the agreed time and to proceed at the agreed placing rate.

Once the pour is under way the contractor should check on the rate of placing and if this is below target the contractor should take corrective action with the placing crew, or in the event of slow concrete deliveries take this up immediately with the concrete supplier. An experienced site supervisor will soon decide by observation whether there are enough people in the placing crew or if they are capable of handling the concrete at the specified delivery rate.

The concrete supplier should also be checking the delivery rate at regular intervals without waiting for prompting or complaints from the site. If serious delays are arising due to an inadequate number of truckmixers, the concrete supplier should take steps to supply extra trucks from other depots. Quality control at the mixing plant should be automatic and at site monitoring the pipeline pressure provides a good indication of concrete quality. The contractor will also normally make arrangements for monitoring the pipeline pressure and for slump testing and cube making which can be arranged so that there is no delay while these operations are carried out. Slump checks which are not carried out until the truckmixer has backed up to the pump will create a delay. Some ready-mixed concrete suppliers will bring their own laboratory vehicle and carry out their own testing — a comforting sight to the contractor and to the pump hire company. It will also impress the resident engineer!

During a pour a tight control on delivery speed and quality control are matters of prime importance which should be given attention alongside the other listed items such as adequate access, availability of stand-by plant, adequate washing out facilities and so on (see section 9.6.). As the pour nears completion certain actions are required if a delay waiting for the final load of concrete is to be avoided. It is a recognized failing in the construction industry that the contractor will wait until the penultimate truck load is placed before calculating the final load quantity. This figure is more often than not taken back to the concrete plant by this truckmixer which then waits to be filled and then returns to

Fig. 100. Two pumps placing 700 m^3 on A55, Colwyn Bay (courtesy Pochin Group plc and Fairclough Civil Engineering Ltd)

the site! Of course, this situation creates a delay of at least half an hour. The delay could be avoided if a final check on quantity is made before the penultimate load is discharged and the quantity is then telephoned immediately to the concrete plant. The figure may not be as accurate, but a small amount of excess concrete is far cheaper than a half hour's delay to the pump and placing crew. The ready-mixed concrete supplier could alleviate the problem by a timely telephone reminder to the site as the pour nears completion. After all, the sooner the supplier is able to finish this job the sooner the next job can be started.

Figure 98 shows a well planned 800 m^3 pour at Runcorn, using four mobile concrete pumps. The concrete had to be placed over a relatively short period of time. Fig. 99 shows a relatively small pumped concrete pour at the UK's first pressurized water reactor nuclear power station at Sizewell, Suffolk. However, on the same project a massive single pour of 4800 m^3 is planned. This pour, which is for a ring round the reactor base, will obviously be given detailed planning and feasibility studies will take into consideration the technicalities (shrinkage and heat of hydration problems), as well as the number of pumps, concrete supply, and so on, which will be required. Fig. 100 shows another well planned concrete pumping operation, in North Wales.

Back-up plant and equipment

The extremes in back-up plant and equipment range vary from the stand-by pump to the humble poker vibrator. It is obviously not commercially sound to provide a stand-by pump except on rare occasions for the larger pours and this is seldom asked for or indeed expected. For a contract which involves more than one pump or a special pour which must be completed within a time limit, a stand-by pump may be requested or may in fact be provided automatically by the pump hire company without any prompting from the contractor. The smaller items of back-up plant can be equally vital in the event of a breakdown (see section 8.3).

8

Common problems on site and corrective measures

The most common problem associated with concrete pumping is a blockage. It is important to be able to recognize a blockage, to be aware of the various causes of blockages and to be able to locate and deal with them promptly. But certain organizational problems which cause delays may also arise on site during a pour.

8.1. Blockages

If the concrete fails to emerge from the end of the pipeline, then provided that the pump is mechanically sound, there is a blockage somewhere in the system. This will be indicated by an increase in the pressure shown on the pressure gauge and by the reaction of the safety valve, which, when activated, causes the pump to cut out. If the pressure rise is rapid with the safety valve acting immediately, the blockage is near the pump. A slower build-up of pressure indicates that the blockage is somewhere in the pipeline, possibly nearer the delivery end. Most blockages occur at tapered sections at the pump end where the concrete deforms.

In the early days of concrete pumping after the Second World War, when the use of this method of concrete placing became more frequent, a blockage was usually greeted on site by such remarks as 'the pump's broken down' or, 'the pump's no good'. Little credence was given to the fact that the mix might not be suitable for pumping and indeed there was considerable ignorance in the construction industry at that time as to the importance of the mix design. The situation is now quite different — concrete technology has developed along with the pumps themselves to provide trouble-free concrete. The mechanics of concrete pumping, however, as described in chapter 4, must be understood and the mix design carried out along those lines. A concrete pump does not change the properties of the concrete, with the possible exception of a loss of air in air-entraining concrete (see section 4.4).

Principal causes of blockages and their characteristics

The three main causes of blockages, in order of significance, are mix design deficiencies; pipeline and joint deficiencies; and operator error or careless use of the hose end.

Mix design deficiencies

Mix design deficiencies become apparent mainly in segregation and bleeding. The mechanics of concrete pumping (see section 4.1) describe how a homogeneous plug of concrete moves along the pipeline, the inside wall of which has a lubricating coat of cement grout. If the water is removed from the plug, the film of grout is lost and a blockage will occur. Water is the only constituent material in concrete which is capable of being pumped. Remove the water and pumping becomes impossible. As the water is forced through and the grout film lost, the aggregate is pressed against the pipe wall, creating excess friction which is sufficient to form a blockage. The dry plug will show the aggregates in their natural state and colour, because the cement grout will have been squeezed out as the blockage is formed.

A further cause of segregation is insufficient mixing. Even a well designed mix with proper grading of aggregates can be affected. Adequate mixing time is important in all concrete and in the case of pumped concrete it is a necessity if the mix is to be successfully pumped. The reasons are clear: the aggregates must have a full coating of cement grout if the required homogeneous mass is to be produced.

Recognizing bleeding presents no great difficulty, as the grout appears watery and is discharged at the hose end together with concrete. In the minutes after the pumping stops, water can be seen building up in the pump hopper. Water may also be observed seeping from pipeline joints.

Consistency of the mix. Pumping is not possible if the mix consistency is either too soft or too stiff. Table 27 illustrates the consistency range from stiff through to soft. The descriptions are in general terms only. At the top of the range the concrete will not pump because it cannot be sucked back into the cylinder and at the bottom end the extreme softness causes segregation and bleeding when pumping is attempted. Provided that the mix design is in fact a pumpable mix,

Table 27. Basic consistency range

Consistency	Spread (Euro): cm	Equivalent slump: mm	Appearance of concrete	
Stiff	— Not measurable	—	Crumbles when poured	Unpumpable
Plastic	36–42	60–100	Surface glistens, firm/soft	↑ Pumpable ↓
Soft	42–50	80–200	Free flowing	Unpumpable

the plastic range will be pumpable as well as the lower end of the stiff range and the higher end of the soft range of flowing concrete. The experienced pump operator knows that if the concrete will go into the cylinder then there is at least a 90% chance of pumping it.

Concrete in these categories therefore can cause blockages, in the valve mechanism if the concrete is too soft and at the pump or immediately behind the pump if it is too stiff, e.g. at the tapered section of the pipeline. Blockages of this type can also be found at bends in the pipeline or in the boom section of a mobile pump.

If the consistency is too soft, bleeding can also occur producing segregation as previously described. A blockage due to consistency extremes is indicated usually by the valves jamming (soft) or sluggish piston action (stiff). Unpumpable mixes will cause a build-up of concrete in the valve mechanism. Symptoms of excess fines can be recognized by a build-up of sand in front of the reduction pieces, because the sand has a higher coefficient of friction than the pipeline and will tend to stick under pressure.

Additives and special cements. If additives and special cements are used incorrectly they may cause blockages. Detailed knowledge of the additive and of the type of cement being used is very important. Reactions may be quite different depending on the temperature and quantities involved. Excess amounts of air-entraining agents can produce a cushion effect, particularly in long pipelines. The pump may be operating satisfactorily but the concrete is not being pushed out at the delivery end: in fact, another type of blockage. The concrete can sometimes be heard moving backwards and forwards in the pipeline and this will most certainly be leading to segregation.

Retarders should also be used with care as an increase in temperature can produce the opposite to the intended effect — they can become accelerators. Any undue delay in concrete supply or in placing can result in the concrete setting when accelerators are incorporated in the mix. Such blockages can create severe problems in clearing them as the whole of the pipeline may be affected with concrete setting rapidly. More often than not it becomes a matter of replacing lengths of pipeline, which is a very costly affair.

High grade cements. These must be used with a great deal of caution and it is recommended that pumping should be continuous when such cements are included in the mix. Periods of delay such as stripping down the pipeline or in blowing out must be kept to a minimum. Large temperature changes will accelerate setting time. Blockages due to early setting can easily occur unless extreme care is taken and only skilled operators should be used.

Pipeline and joint deficiencies

A dirty pipeline which has resulted from inadequate cleaning out, will cause blockages which normally occur at the point in the line where the concrete has set. On the other hand, a generally dirty pipeline can inhibit the lubricating grout flow which again will create a blockage from bleeding and segregation. Clean pipelines are therefore essential: an obstruction in the pipeline can cause a blockage.

Worn or dirty hoses are a potential danger. Before the start of any concrete pumping pour they should be checked by the foreman or supervisor. A blockage can easily be created in a hose which has not been cleaned out properly after the previous pour. Weakened worn areas will rupture more easily. The initial pumping may often reveal a blockage in the end hose. It is worthwhile to carry out a thorough visual inspection of the hose — after all, it is possible to look through it — and in very hot weather to flush through with water before the lubricating mix is put through the pipeline.

Bends incorporated in a pipeline which are too sharp or too numerous increase the concrete pressure to such an extent that a blockage results.

A blockage can easily occur in a sloping pipeline (downhill) due to gravitional effects on the lubricating mix and on the concrete. The procedures to be adopted to counteract this are described in chapter 9.

Poor jointing resulting from either inadequate laying of the pipeline or from worn or damaged joints can result in loss of grout and lead to subsequent blockages. With long pipelines it must be recognized that a sufficient quantity of lubricating grout to cover the complete length of pipeline has to be pumped through before pumping the concrete.

Operator error or careless use of the hose end

If the operator has failed to lubricate the pipeline properly, including the hose end (see section 9.3), the resulting increase in friction could be sufficient to create a blockage. This may also inhibit pumping by operating at too high a pressure, or, in other words, overpumping. A mix may be quite suitable for pumping but may not pump for this reason. Pumping a sufficient quantity of grout through the pipeline is essential (see section 9.3).

Extreme care should be taken in handling the flexible rubber end hose as careless handling, e.g. kinking, can cause a blockage. An inexperienced placing crew mishandling the end hose can create many problems, not least blockages.

Even though these are the main causes of blockages, there can of course be faults in the pump itself (however rare), which prevent the concrete moving. Such occurrences may not strictly be classed as blockages, but a pump which is in poor condition with worn cylinders, pistons or valves could be a problem. The pumping pressure available may be insufficient to move the concrete over the full length of the pipeline or because of the nature of the mix.

8.2. Clearing blockages

A minor blockage may be cleared by forward and reverse pumping, but no amount of pressure will clear a major blockage. Excess pressure will only make matters worse under certain conditions and could result in a serious accident.

Sometimes shortening the pipeline will reduce the pressure and on restarting pumping the blockage will clear itself.

Tapping the pipeline with a hammer can often locate a blockage. When the line is full of concrete under pressure, a hammer tap will produce a thud, but the pipe will ring where there is no pressure. The blockage will be at such a junction.

Before attempting to dismantle any part of the pump or pipeline to clear a blockage (or for any other reason for that matter), the pressure must be relieved by reverse pumping. It should be remembered that reverse pumping may not relieve all the pressure, particularly when the pipeline or boom is inclined downhill.

A blockage should be cleaned out by rodding or by using a sponge ball (provided for normal cleaning out operations) with compressed air or high pressure water. Appropriate safety precautions must be taken (see section 12.7). If this does not remove the blockage the pipe(s) should be replaced, but new lengths of piping must be thoroughly lubricated before restarting pumping.

The taper piece immediately behind the pump is the most likely place to look for a blockage in a modern concrete pump. Normally these taper pieces can be dismantled easily and the blockage removed quickly. If no blockage is found on dismantling the taper piece, the pump should be restarted to see if the blockage is located at the valve mechanism. This should be done before reconnecting the taper piece.

If a concrete pump is operating properly when idling, the blockage cannot be attributed to the pump and must be due to a concrete deficiency. Therefore time can be saved by checking the mix first.

If the valve mechanism is not working properly, a blockage has occurred. Precautions must be taken to isolate the concrete valve by closing the shut-off valve before attempting to clear any blockage in the area of the concrete valve or pistons.

Weather conditions should always be borne in mind. In very hot weather concrete will set quicker and even a short delay could present a problem. Similarly, certain admixtures may cause early setting (see section 4.4).

Placing crews should stand well clear of the delivery hose until a blockage is cleared, as the hose may whiplash when an obstruction is cleared.

8.3. Organizational problems on site causing delays

Many contractors consider that, once they have prepared the pour efficiently and are ready at the time agreed with the concrete supplier, there should be no further organizational problems during the pour. There are, however, several other problems which continually crop up. The preparatory work for the pour may be of a high standard, but sometimes little thought is given to ensuring good continuity during the pour.

Access

The necessity of providing and maintaining good access is appreciated by most contractors, yet on several of the pours observed in the Lancaster University research,[1] inadequate access caused significant delays. Suitable access for normal site operations might not in reality be suitable for concrete pumping, because full continuity must be maintained during the latter to exploit the full potential of this method of placing concrete. A case study in the Lancaster University research report illustrates what constitutes proper access for the pour (Fig. 101). The contractor realized that much time can be wasted in manoeuvring truckmixers on site and although there was insufficient room to provide an access road that was wider than a single truck, waiting areas were provided. With an uninterrupted supply of truckmixers, particularly on a large pour, the provision of waiting areas is essential to cope with the unavoidable build up of vehicles from time to time. In the illustration, there was no room for more than one truckmixer at the pump position, yet there was very little delay in the next truck backing up, as shown. Without the waiting areas there could have been occasions when there was a queue of truckmixers on the main road, not only causing traffic problems but creating delays in getting the truckmixers up to the pump quickly. On large pours particularly (but at the same time on any pour), a few minutes lost in continuity of supply soon adds up to a significant amount of time. The placing rate of this pour reached nearly 50 m^3/h, more than double the average rate for a single pump of the size used. Outputs on this site regularly reached

Fig. 101. Maintaining access for a pumped pour: foundations for the Potteries Centre (courtesy Shepherd Construction Ltd)

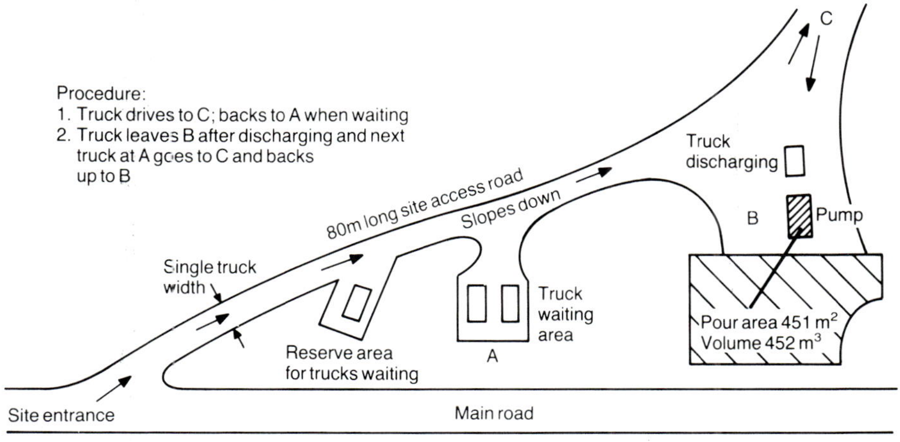

Fig. 102. Pumping concrete into columns: a smaller pour at the Potteries Centre (courtesy Shepherd Construction Ltd)

such a figure, showing what can be done with good planning (Fig. 102).

Restricted access roads

On some sites it may be impossible to provide more than a narrow access road, and passing places or parking areas cannot be constructed simply because there is no space available. The longer the access road, the bigger the problem becomes. In concrete pumping, when continuity of work is essential, the problem becomes more acute. A load of bricks being held up is likely to be less serious or costly than a truckmixer being held up in the middle of a concrete pour, when all the associated plant and labour would also be affected. Some site organizations fail to recognize this potential problem and deliveries to site continue with no thought having been given to blockage of access for the concrete supply. In most cases a few telephone calls from the site to the materials suppliers to bring forward or delay deliveries so that the pour is not interrupted would neither affect the contract as a whole nor the concrete pour in particular.

Several instances of access delays which could have been avoided, were recorded during the Lancaster University study and there is no question that this type of delay is significant and costly.

Condition of access roads and hardstandings

The condition of access roads can lead to delays. The responsibility of providing and maintaining an adequately surfaced access road or hardstanding for the pump normally rests with the contractor. Sometimes the importance of this is overlooked and a single bogged-down truckmixer can seriously affect the continuity of the work. The site agent should bear in mind that a fully laden 6 m^3 truckmixer weighs about 24 t, distributed on perhaps three axles. The pump position should be adequately prepared so that the ground will easily support the outriggers. It should be remembered that the pumps are designed to operate with a full load of concrete in the hopper and boom (in the case of the mobile pump), but this is dependent on the ground being capable of supporting such loads (see section 12.5). The other requirement at the pump position is for this hardstanding to be level and the area for the truckmixers backing up to the pump to be level as well.

Most truckmixers carry towing chains — not principally in case of a breakdown, but because they may be bogged down on site! This occurs frequently in winter. Protestations from the truck driver when asked to drive over a doubtful area are sometimes met by an assurance from the site foreman that there is an excavator around which can pull the truckmixer out! Other truckmixer drivers will say that they are reluctant to enter certain sites unless they can see that there is an item of plant there which is capable of coming to the rescue! Not only does it make commercial sense to provide good access, but also the importance of safety must not be overlooked. Unfortunately there have been many instances of pumps and truckmixers overturning because of ill-prepared surfaces.

Continuity for the pour

Communication between the parties to a concrete pour is described in chapter 7. Good communications between the pump operator and the placing crew are also essential. With extended pipelines a clear system of signalling must be arranged between the pump operator and a previously appointed signalman at the placing end. They may be far enough apart to inhibit visual signalling and there may be occasions when the pump operator cannot see the placing position. In these circumstances short-wave radios are useful or a second signaller in view of the pump operator should be employed to relay signals from the pour position. Hand signals should be worked out carefully and rehearsed. After all, the pump operator may have to stop pumping suddenly and the signal for this has to be clear.

Poor communication on site can adversely affect continuous working. A stop–start situation can develop which usually worsens as the pour proceeds. The placing crew may become discouraged if they cannot get on with their work steadily and retrieving such a situation is extremely difficult.

Stand-by equipment, such as vibrators and screeders are low cost items, yet the whole pour can be stopped if a breakdown occurs to one of these items and there is no spare on site. Instances do occur when, for example, there is no spare vibrator on site and the one being used breaks down. It can take over an hour sometimes to locate another vibrator and bring it to the site. Comparing the hourly cost of a placing crew, pump operator and pump (not to mention the delay cost incurred by the concrete supplier) with the hire cost of small concrete vibrator, even on a weekly basis, shows that there is a good deal of sense in ensuring that stand-by equipment is available at all times.

Interruptions for meal breaks need not occur provided that some simple arrangements are made for members of the placing crew to take meal breaks at different times. While two or three members of the crew take a meal break, the others can continue with numbers made up from other operatives on site who may be made available from less important work. The whole of the crew might not then be fully experienced in such a situation, but at least the pour is continuing, which is most important.

9

Hints for site operatives

The pump operator should be aware of what constitutes a well laid-out pipeline and ensure that it is properly lubricated before pumping the first load of concrete. The pump operator should also know how to pump concrete in various situations such as downhill or in extreme weather conditions. An experienced placing crew on the other hand will know how to use the end hose safely and to the best advantage. Training is therefore essential for all site operatives. Care of equipment and minimizing wear and tear are extremely important.

9.1. Laying the pipeline

The concrete pump should be positioned on site so that the pipeline will be as short as possible. By doing this there will be a saving on the length of pipeline used. Labour costs in laying will be reduced as well as in handling and dismantling. Furthermore, the shorter the line the lower is the pressure required which in turn reduces the wear and tear on the pump and pipeline. Any change in direction means using a bend, which again increases pressure. Therefore the pipeline should be kept as straight as possible by using the most direct route available from the pump to the placing position.

Careful planning of all aspects of laying a pipeline, i.e. diameter to be used, direction, access to the line, anchoring arrangements, special attachments required and most certainly ensuring that the pipes and couplings are in good condition, will go a long way towards a successful concrete pumping operation. The importance of using the correct wall thicknesses for pipes, correct type of couplings, etc., have already been described in section 2.2, where Table 3 and Fig. 28 indicate safe working pressures and wall thicknesses for various pipe diameters.

Long pipelines

There is considerable movement of the pipes, particularly with longer pipelines, irrespective of whether they are laid horizontally or vertically. This is due to the pulsations created by the pump action: these pulsations increase with the length of line. It is evident that the pressure will decrease at the end of the pipeline, but over the whole line, the longer the pipeline the greater the movement.

It is necessary to have some degree of flexibility in a pipeline and the couplings used should provide for a small angular movement between the pipes they join. The pipes, however, must be securely anchored and adequately supported, yet at the same time be free of tension.

Anchoring and supporting pipelines

There are certain parts of any pipeline where forces are created due to directional changes, e.g. at bends, and at these positions there is likely to be considerably more movement. It is here that special care must be taken and firm anchoring is essential. A horizontal pipeline should be simply supported on wood blocks, with care being taken to keep the line off formwork or reinforcing steel (Fig. 103).

Clamps should be used at bends in the line. These can be made in various forms, usually bolted to concrete blocks or attached to parts of the structure. Before anchoring the pipeline to any part of a structure, falsework or scaffolding, agreement and approval should be obtained from the main contractor/client as to the adequacy of the anchoring point and the method of anchoring. Fig. 104 shows neat steel clamp arrangements at bend positions on the ground, and Fig. 105 gives two methods of supporting a bend at the horizontal and vertical junction. This particular support must be designed for dealing with the transition forces between the vertical and horizontal and ensuring that the bend is not carrying the full weight of the vertical pipeline, which should have its own supports for dealing adequately with gravitational forces as well as the weight of the pipeline.

Threading pipelines vertically through buildings offers

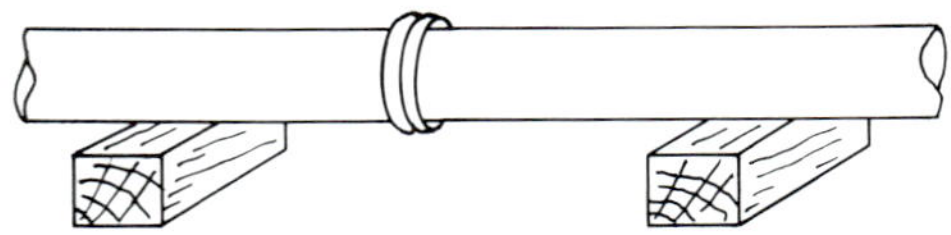

Fig. 103. Simply supporting a pipeline on wood blocks

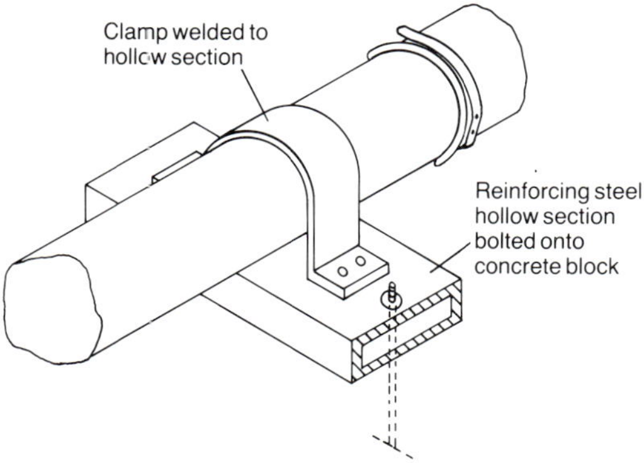

Fig. 104. Clamping the pipeline at a bend

scope for ingenuity in making use of stairwells, lift shafts, or other appropriate access positions in the structure. This can be particularly useful in refurbishment work. The pipeline can be chocked or wedged securely through available apertures.

But a pipeline is not always laid on the ground or up a building. Fig. 106 illustrates one novel method of laying a pipeline across a lake. Floating oil drums provided the necessary support.

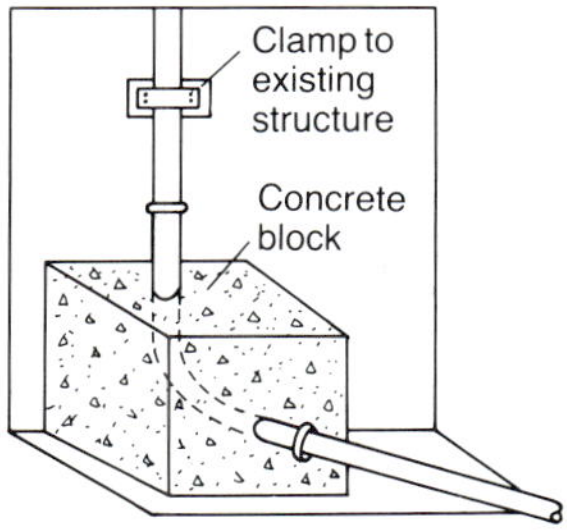

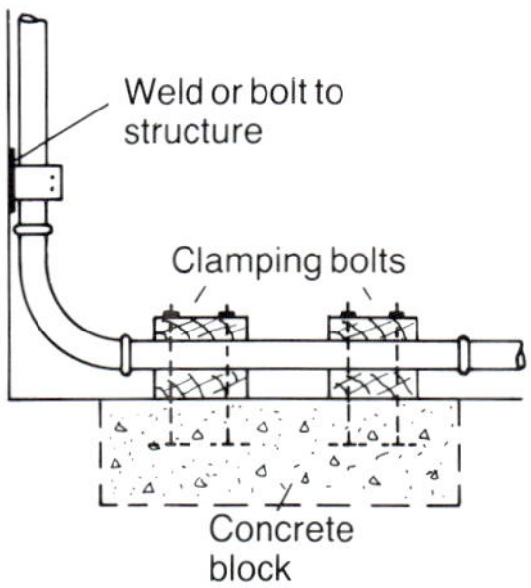

Fig. 105. Supporting a bend at horizontal and vertical junctions

Fig. 106. Supporting a pipeline across water (courtesy Pochin Group plc)

Horizontal distance for vertical rise

Gravitational forces dictate that certain procedures should be adopted when positioning a pump which has to feed a vertical pipeline. It is obvious that the concrete in the horizontal section of the pipeline will act as a counterweight to the concrete in the vertical section and that it becomes increasingly important the longer the vertical section of pipeline is. Therefore a horizontal starting distance must be determined which is related to the pumping height and the pipe diameter. It is not possible to be categoric about such starting distances as each situation will be different. The type of pump and output capabilities must also be taken into account. A simple rule of thumb which has proved satisfactory in the past is that the starting distance should be roughly 10–15% of the vertical distance (Fig.107). Obviously in the case of a pump conveying concrete to several floors of a building, the highest vertical height must be used in the calculation.

It is of great importance that such starting lengths are anchored firmly. It is an overworked length and should be treated as such. Wear on the bend will be high and any blockages are likely to be on this horizontal length of pipe. For this reason alone, a short, say 1 m, pipe section, which can be handled easily, should be included in this starting length.

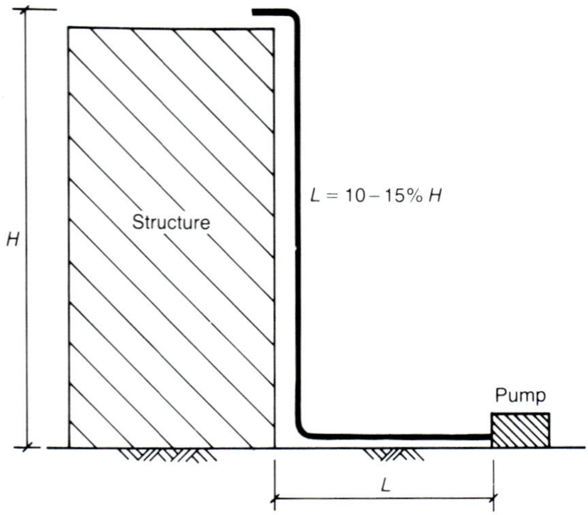

Fig. 107. Horizontal starting distance

Sloping or inclined pipelines

When pumping through sloping or inclined pipelines problems arise because it is necessary to prevent the concrete from running away due to gravity. The problem becomes more acute if there is very little horizontal line at the bottom of the sloping section. Concrete will flow by itself down a descending pipeline; the rate of flow is governed by the fluidity of the mix, the pipeline diameter and the degree of inclination. Segregation will occur. It is clear that to maintain a full column of concrete in the descending pipeline, the capacity of the pump must be sufficient to achieve this. A facility must also be provided for retaining the column of concrete. At one time stowing bends or S type bends were used to act as brakes, but these are now seldom used. The answer to the problem is normally to incorporate a shut-off valve at the bottom of the descending line, thus making it possible to maintain a full concrete column if, for any reason, pumping has to stop. These hydraulic shut-off valves (Fig. 108) can be operated by remote control by the pump operator, who has a warning light to indicate whether the valve is open or shut.

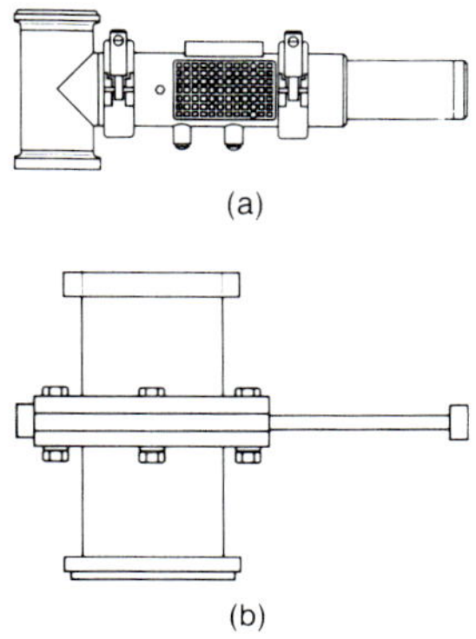

Fig. 108. Shut-off valves

When lubricating a sloping pipeline with grout it is necessary to ensure that the whole of the inner surface of the pipe is lubricated. In this case the answer is to insert a sponge rubber ball in front of the grout mix (this is normally not a requirement with horizontal pipelines). As with horizontal and vertical pipelines, sloping pipelines should be anchored properly, and should also have some flexibility. By using a shorter length of pipe just above the bottom end of the descending line, cleaning is made easier because the lighter pipe can be removed easily. Descending pipelines should not be laid transversely.

9.2. Siting the concrete pump

If maximum outputs are to be achieved, and ready-mixed concrete is being used, there should, if possible, be sufficient space at the rear of the pump to accommodate two truck-mixers (Fig. 109). On high output large pours, the time taken for one truckmixer to move away from the pump and for another to reverse up to the pump can be significant when added up over a full day. With a second truckmixer in position ready to start pouring concrete into the pump hopper the moment the first truckmixer is empty, no time at all is lost.

Fig. 109. Two truckmixers at the concrete pump hopper

A mobile pump used for a specific pour need not have the same installation requirements as say a large stationary pump which may be programmed to remain in one position for several months. In both cases, however, the general requirement of firm, level positions is of course absolutely essential, but when installing a stationary pump additional factors have to be taken into account.

The pump is going to be in that position for a long time and it has to be cleaned out and maintained regularly. Therefore placing it on a concrete slab with drainage facilities is more sensible than using a few timber baulks. In addition, access around the pump is a prerequisite for proper maintenance, repairs and cleaning. Sufficient room should be provided for the plant fitters and the pump operator to carry out this work. The work will be done more quickly and efficiently if there is sufficient room. Spending some money on such facilities at the outset will often save a great deal more at a later stage. The pump should be sited so that the first bend is well clear of the pump — normally at least 10 m away.

9.3. Commissioning the concrete pump

As described in chapter 4, it is necessary to lubricate the pipeline with a cement/water or cement/sand/water mix, before the first load of concrete is pumped. The former mix is used with short pipelines, e.g. mobile pumps with placing booms. For pipelines over 50 m long it is usual to include sand in the lubricating mix. The consistency is soft. Normally the pump operator will mix the grout by hand in a drum and insert it into the line just ahead of the pump.

Quantity of grout required

For mobile pumps with booms up to 25 m long, 50 kg of cement/water is usually sufficient. Similar pumps with booms up to 30 m long and 125 mm dia. delivery line will require at least 100 kg of the same mix. Schwing have compiled a table for pipelines of different sizes and lengths indicating the quantities required. The size of sand is limited to 4 mm (Table 28).

It is usual to wait until the first load of concrete has arrived at the pump before grouting the pipeline. The mix should be pumped through the pipeline slowly and smoothly. Care should be taken to discharge the mix in an area away from the actual pour, preferably somewhere which has been agreed with the site foreman. If the weather is very hot it is advisable to wet the inside of the pipeline with water. This can be done by placing about 50 kg of water in front of the lubricating mix and separating it from the mix with a sponge ball. If the pipeline is particularly long it is recommended that a plug of water positioned between two sponge balls is pushed through the line using compressed air. Immediately afterwards, the lubricating mix is pumped through followed by the first load of concrete. Using the water plug is one means of checking if there are any leaks in the pipeline and indeed if it is free from obstructions.

With descending pipelines, a rubber ball is used as previously described but, as the difference in level between pump and placing position increases, it is sometimes neces-

Table 28. Minimum quantities of lubricating grout for pipelines over 50 m long

Length of pipeline: m	100 mm, 125 mm dia. pipeline			150 mm, 180 mm dia. pipeline		
	Quantity of lubricating grout: l	Cement: kg	Sand: kg	Quantity of lubricating grout: l	Cement: kg	Sand: kg
50	200	200	100	300	300	150
100	300	300	150	400	400	200
200	500	500	250	750	750	350
300	750	750	350	1000	1000	500
400	1000	1000	500	1500	1500	750
600	1500	1500	750	2000	2000	1000

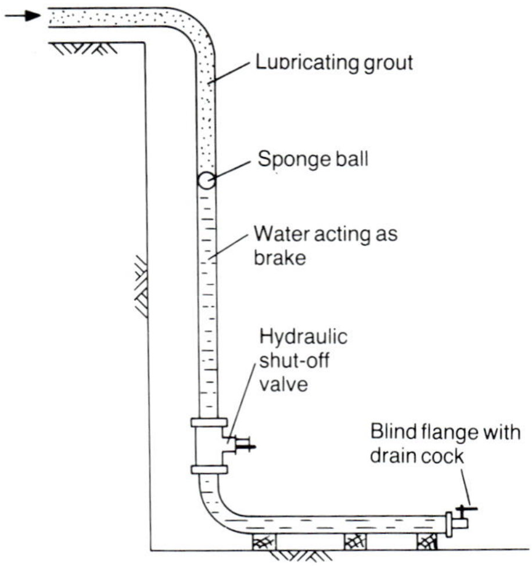

Fig. 110. Lubricating a descending pipeline

sary to insert water first in the line to act as a brake and to control the descending lubricating mix. The water and the lubricating mix have to be separated by a hard rubber ball, wad or watertight plug. The arrangement is shown in Fig. 110. The line is filled with water, a blank flange complete with drain cock being fitted to the discharge end. The rate of descent of the lubricating mix is controlled by operating the drain cock, which in turn is adjusted to the pump pressure. This arrangement is also very suitable for cleaning the pipeline using water. Before starting to pump the concrete, the blind flange and drain cock are removed.

Pumping the first load

The pump operator should have been thoroughly trained in all aspects of this work. Before pumping the first load the pump operator will have checked that the pump is functioning properly and that the pipeline has been properly laid and does not leak. Once the operator is satisfied with the adequacy of the lubricating grout, the first load of concrete can be pumped. It should always be remembered that even the most suitable mix will not pump if there is an inadequacy in the pump and/or pipeline.

An experienced operator will inspect the concrete visually as it comes down the truckmixer chute and check that the concrete looks pumpable. Unpumpable mixes will look dry and lumpy. Concrete which is breaking up as it drops off the truckmixer chute is not really suitable for pumping.

After the operator has checked that the hopper grille is in position and that the concrete has been thoroughly mixed in the truckmixer, the concrete can now be fed into the pump hopper. Observing the nature of the concrete material carefully at the start and finish of the discharge of a full load from a truckmixer can provide useful information on the quality. Pumping should proceed with caution as an apparently suitable concrete entering the hopper may misbehave during pumping and a blockage could occur. Applying too much power at the start of pumping can cause bleeding and segregation. Therefore the operator should check that the mix is pumping satisfactorily in all respects before increasing output speeds. In nearly every case, the pump capabilities will far exceed that of the placing crew and there is no point in pumping at higher pressures than actually required. In the same context it is worth remembering that the higher the pressure the more wear there will be to the pump and pipeline. Should there be any problems with the first load, approval must always be obtained from the site management before any changes are made to the mix, e.g. by adding water or fines. If the concrete appears to be pumpable but there is difficulty in pumping, the speed of pumping should be reduced as a first step. The problem may be the pipeline: it could be too long, there could be too much flexible hose, or the joints could be leaking. Even after careful preparation of the pump and pipeline as indicated previously, a mix may not pump. It is then preferable to redesign the mix than to continue trying to pump, which will probably result in blockages and more delays.

9.4. Placing concrete in extreme weather conditions

Concrete pumps have been used successfully at temperatures between −20°C and +40°C. It is not recommended, however, that concrete pumping at temperatures anywhere near these extremes is to be encouraged. The first consideration concerns the pumps and the necessity to comply with any rules and regulations laid down by the manufacturers in relation to special hydraulic or motor oils to be used when operating under any extreme conditions. The effects of extremes of heat or cold on the concrete is perhaps the foremost consideration, but it is equally important that the pumps are capable of operating in these conditions. Some pumps are equipped with oil coolers for work in high temperatures.

Pumping concrete at high temperatures

In hot countries certain steps are taken to enable concrete to be pumped successfully. The obvious steps include preparing the concrete with cooled aggregates and cooled water and to avoid any interruptions during the work. Pours may be undertaken either in the early morning or in the evening. Recently large pours have been carried out successfully in the Middle East from early evening through to early morning.

As far as mobile pumps are concerned no special precautions are necessary apart from those which have already been mentioned. Special precautions are required when using stationary pumps with perhaps long pipelines. Warm concrete sets faster and the lubricating film will be lost from the pipeline wall more readily when concrete is pumped through a hot pipeline. The concrete will become tacky and there will be an increase in friction. This in turn raises the possibility of bleeding and blockages. Protection of long pipelines is therefore desirable. This can be done by covering the pipeline with something which is kept damp. Even painting the pipeline white can be beneficial.

Before pumping the lubricating grout through the hot pipeline, the hot pipeline should be cooled with a hose pipe and thoroughly soaked through inside with water. The lubricating mix should be thinner than normal. The pipeline joints should be as tight as possible, to minimize the grout loss.

Pumping concrete at low temperatures

Again, the manufacturer's instructions must be strictly adhered to with respect to oil viscosity requirements, and also with regard to starting up in cold conditions. It is possible to pump concrete successfully if the concrete is prepared from heated aggregates or hot mixing water, but an adequate mixing time is most important if bleeding is to be prevented. After hot concrete has cooled, it has been found that bleeding sometimes occurs.

As with warm concrete in high temperatures the concrete sets quickly, in this case from using pre-heated aggregates. Again, interruptions must be kept to a minimum and the pipeline should be cleaned immediately on completion of the pour. The pipeline should be opened at low points to release all water which might otherwise freeze and cause a blockage the next time it is used.

9.5. Use of end hose

The distribution or end hose attached to the end of the pipeline or boom (in the case of mobile pumps) is an extremely important part of the pipeline and indeed of concrete pumping generally. Yet on many construction sites its use is given scant consideration compared with the attention given to other matters which make it possible to pump the concrete out of it. The tendency is to think that once the concrete is discharging nicely from the end hose then that is the last of the worries. One can almost hear such remarks as 'any fool who is strong enough can pull a hose around'! However, a check-list concerning this part of the concrete pumping operation very quickly reveals its importance.

(*a*) Is the correct type of hose being used? For example, has the hose been marked with the safe working pressure?
(*b*) The length should be limited and should not exceed 3 m for 150 mm dia. pipeline and 4 m for 125 mm dia. pipeline.
(*c*) Is the end hose firmly attached to the boom end or pipeline end?
(*d*) Has the end hose been thoroughly checked for cleanliness, wear, bulges?
(*e*) Has the handler been properly trained in the use of the hose?

These five simple points may not appear to be significant at first glance, but in reality all are extremely important. To disregard any is taking a chance which could lead to a serious accident.

Using the correct type of hose recommended by the manufacturer for a specific pump is very important. Marking it for identification purposes makes sense. In this way it is unlikely that an inferior hose which is unable to withstand the pressure from the pump can be fitted. It might be suggested that the concrete pressure at the placement end is zero, but a blockage, for whatever reason, could occur in the hose itself, with disastrous results.

The length of hose should be limited for several reasons. It is very difficult to handle more than 4 m of 125 mm dia. hose full of concrete. 3 m of 150 mm dia. hose will weigh about 170 kg. In addition, the longer the hose the more likelihood there is of kinking. Kinking is extremely dangerous and the resulting build-up of pressure will create a form of blockage or cause the hose or the connection to the pipeline to burst. Whiplash and concrete erupting at high pressure can cause serious accidents (see chapter 12). A 3 m long end hose can only be moved 2 m each way before the end hose starts to sink into the concrete.

The length of end hose coupled to the boom of a mobile pump is particularly critical. An excessive length of hose full of concrete can in certain conditions create bending moments which are not catered for in the design of the pump. It is not practical to include these moments in the stability calculation for such conditions. Even the uninitiated should realized that an extra weight dangling and moving in all directions from the end of a boom can produce forces which could cause the boom to collapse or the pump unit to overturn. Manufacturers design conservatively and with factors of safety, but it is necessary to adhere strictly to all their rules and regulations. They are provided for the benefit and safety of the users.

Firm connections or the use of proper couplings between the hose and the steel pipeline are clearly necessary. A kinked hose or blockage in the hose section will produce a build up of pressure and the most likely breakage point is at the connection. Furthermore, the hose suspended from a boom must always be a hazard, particularly if the joint is not in good condition.

The importance of training the hose handler adequately cannot be emphasized enough. The observer of a concrete pump pour can easily appreciate or recognize a well-trained hose handler, by the way in which the hose is manipulated with an apparent minimum of effort, where the handler starts to place the concrete (working back systematically), the communication signals and so on. The handler with the hose slung over a shoulder and staggering about unable to keep his or her balance is a danger to him- or herself and the rest of the placing crew. The handler may also have one or two colleagues pulling on rope slings looped round the hose, which is probably much too long in the first place and is why slings are being used. Special care must be taken to ensure that the hose does not cause damage to the formwork or reinforcing steel.

9.6. Correct labour for the job

There is no such thing as a fixed number of people in the placing crew. Each job is different and may require different numbers of people in the placing gang. The correct labour for the pour might be defined as

> 'the minimum number of trained operatives, (shovellers, screeders, vibrator operators, finishers, etc.) required to place the concrete properly at a speed which matches the pump output, when the pump is operating conservatively.'

Put simply this means that the labour should be able to cope with a reasonable output from the pump yet at the same time place the concrete correctly. The permutations of numbers in the placing crew are practically endless and depend on the type of pour, e.g. thin or thick slab, etc. (see section 7.1).

Planning for the requisite number of people is essential if the cost of the pour is to be kept to a minimum. The University of Lancaster Report[1] evidenced serious delays occurring on pumped pours because of a lack of people. Pumping had to be suspended while the depleted placing crew caught up with levelling and screeding. During this time the queue of ready-mixed concrete trucks grew and other customers received their concrete late that day. Too many people may be better than too few, but the extra wages have still to be paid. In addition, a study by Dressel[62] in West Germany in recent years into the behaviour of labour in the construction industry proved that a crew with too many people could lead to a situation whereby labour productivity was actually reduced and there was a detrimental effect.

On some large projects a fixed number of people in a concrete placing crew are employed solely on concrete placing. This can have its benefits as the crew will be experienced and be well versed in pour routines. But with the variety of types of pours which inevitably occur on any project there is some argument as to whether it is a good thing or not to employ such gangs. On many occasions overmanning was clearly evident on pours observed during the Lancaster University study, both in the UK and in West Germany. Each pour should be considered separately and the majority of construction managers are quite capable of making an accurate assessment of the number of people required for a specific pour. It is surely a matter of planning. The number of people could be as little as three or as high as twelve. The Lancaster University report noted that a placing crew of twelve people was unable to place concrete at more than 20 m^3/h for a bridge deck slab, yet a crew of five placed concrete on a large foundation slab at nearly 50 m^3/h.

Having the correct number of people with the requisite experience and skills for the work will without doubt pay dividends to the contractor.

9.7. Cleaning down and washing out

Keeping the pump and equipment clean is an important part of the pump operator's duties. The need to proceed with caution when cleaning out pipelines cannot be emphasized enough. As far as possible, water should always be used and these operations should only be carried out at a position on site agreed with the site management. Even then care should be taken to avoid nuisance by splashing structures, personnel or other vehicles. Care should also be taken to ensure that no cement grout or slurry enters a sewage system. Remedial costs in such an event can be astronomic.

Concrete will set quickly in warm weather and any pipeline lengths removed during the pour should be flushed through with water as soon as possible. If pumping has to be suspended for any reason and the delay is long enough to raise doubts as to the state of the concrete in the pipeline or boom, it is wise to check if there are any signs of setting. This can be recognized when the concrete emerges in dry looking slugs and the pumping pressure is unusually high. The pipeline should be washed out immediately if this occurs. If there is any doubt, wash the pipeline out.

Cleaning the pipeline

Using high pressure water. The preference for using water instead of compressed air is because there are fewer dangers using the former, but sometimes it is necessary to use compressed air, particularly for long pipelines. As with pumping concrete the pipeline is under pressure although in this case water is being used to force a sponge ball or wad through the pipeline. The immediate source of danger is the ball or wad discharging violently from the end of the pipeline. The first requirement is for a specially made wash-out adaptor to be fitted and a catch basket attached at the end of the pipeline (Figs 111 and 112) to retain the wad or ball which may be ejected with considerble force. Concrete is also going to be expelled from the line and should be directed to a safe area.

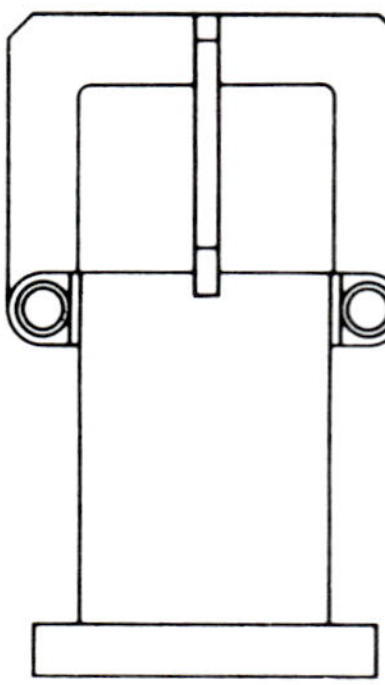

Fig. 111. Catch basket

It is usual to remove the placement hose and any other

flexible hose before washing out, but the static line should remain supported and anchored. Any bends near the end of the pipeline should be removed first before starting cleaning. Personnel involved in cleaning a pipeline should wear protective clothing, eye protection and a hard hat. All personnel should stand well clear of the operation, particularly near the discharge end. To clean out a mobile pump with water, a sponge ball is sucked back from the discharge end to the pump hopper. The pump operator has to wash out the hopper, valve mechanism, and tapered or Y pipes thoroughly. Before working in the hopper the operator must do three things: switch off the engine, vent the hydraulic pressure and put the remix control into neutral and lock it.

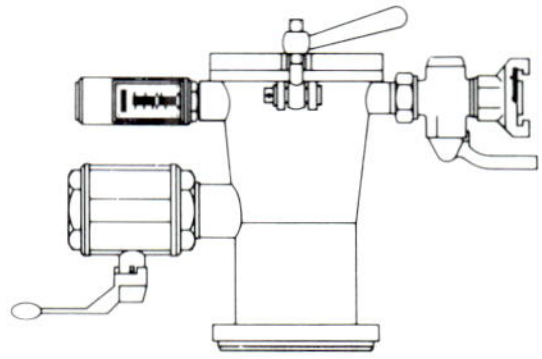

Fig. 112. Wash-out adaptor/pipe cleaning head

Using compressed air. Using compressed air for cleaning is a potentially dangerous operation. The BCPA state in section 6 of the *Manual and advisory safety code of practice for concrete pumping*[6]

> '*CLEANING WITH COMPRESSED AIR* — THIS OPERATION IS POTENTIALLY HAZARDOUS AND SHOULD ONLY BE USED TO CLEAN PIPES AND HOSES USED IN CONCRETE PUMPING WHERE THERE IS NO PRACTICAL ALTERNATIVE, IF THIS IS THE CASE. THE OPERATION SHOULD BE CARRIED OUT UNDER VERY CLOSE SUPERVISION OF A COMPETENT PERSON'.

All the other precautions stated as necessary when using high-pressure water are applicable when using compressed air. In particular, all people in the area should be informed of the operation and the danger involved and instructed to keep clear of the whole pipeline. The operation should only be carried out under the supervision of a competent person who has been fully trained in this method of cleaning out a concrete pump pipeline. The competent person should have a person who is experienced in the operation to assist.

The pipeline may be cleaned out from either the pump end or the delivery end, whichever is the more convenient for depositing the concrete or obtaining the water or air supply. The length of pipe nearest to the pump is first removed and cleaned manually. It is advisable that there are no bends in the last 15 m of pipeline being cleaned.

A cleaning plug (sponge ball or jute bag) is inserted and then a pipe cleaning head with connections for compressed air and water is fitted to the end of the pipeline (Fig. 113). The entry air cock is operated carefully to control the pressure build-up and the movement of the ball or wad along

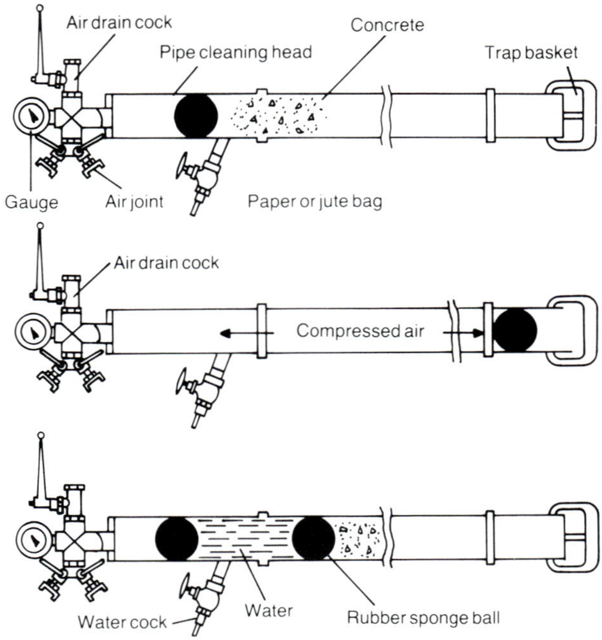

Fig. 113. Cleaning the pipeline with compressed air and flushing out with water

the line. The assistant should check the progress of the sponge ball closely by tapping the line, as the air must be turned off before the ball reaches the end of the line, otherwise there could be a violent discharge. If the speed of discharge of the concrete is too rapid at any time, air must be vented to reduce this to acceptable levels. When the ball is within 25 m of the end of the pipeline, the air should be turned off and vented from the release cock. The ball is accelerating as it nears the discharge point because the pressure is decreasing. When the sponge ball or plug has been safely trapped in the catch basket and this part of the cleaning operation completed, it is usual to clean the pipeline again with compressed air, using two sponge balls with a water cushion of approximately 50 l between them.

The maximum output of the compressor should not exceed that necessary to complete the operation. Using a very large compressor is asking for unnecessary trouble. Normally a compressor with an output of 7 bar will do the job satisfactorily.

It is possible that the line may have to be uncoupled in the vicinity of the sponge ball if the ball becomes wedged during cleaning. When using compressed air it is possible that the air can overtake the ball, forming a pocket of compressed air between the ball and the concrete. Uncoupling should be carried out very carefully and a helmet and face visor at least should be worn for protection. Many accidents have occurred when cleaning pipelines, the majority when using compressed air. These accidents have nearly always been caused by lack of experience and not by failure of the equipment. A golden rule, therefore, is that the process must be controlled by a supervisor who has had thorough training in this method of cleaning. The supervisor must also have an experienced assistant, as one person alone cannot do the job safely and effectively.

Cleaning equipment. The equipment used at the placing position, e.g. screeders, vibrators, must also be washed immediately the pour is finished. Washing such items is always made easier if the equipment is sprayed with mould oil or something similar before the start of concreting.

Some truckmixer drivers may ask the pump operator if they can wash the truckmixer out and discharge the leavings into the pump hopper. This should never be allowed under any circumstances as lumps of hardened material could damage the pump mechanism when pumping restarts, or blockages could be created in the pipeline. The truckmixer drivers should have been allocated their own wash-out area on site by the site management.

Plant and equipment associated with concrete will always accumulate cement grout or even layers of hardened concrete. Build-up can be rapid in corners or other areas where excess material can be trapped easily. Washing equipment frequently and particularly immediately on completion of the current job — not necessarily waiting until the end of the day — is sensible. It should be remembered that it may only take one instance of neglect to create a major cleaning problem or to have a piece of equipment which is found to be inoperable the next time it is required for use: use the equipment, then clean it immediately.

9.8. Minimizing wear and tear

Concrete pumps and equipment are very susceptible to wear and tear, because concrete is an abrasive material. A sensible approach towards minimizing wear on valves, cylinders and pipelines is never pump faster than is absolutely necessary. Use a pump with an output rating well in excess of the rate actually required. For example, if a placing rate of 30 m^3/h is required, use a pump which is rated at 45 or even 60 m^3/h. The faster the pumping is the greater the wear is on parts in contact with the concrete. The philosophy of using plant bigger than the calculated requirements was widespread in the construction industry during the 1950s and 1960s. Despite the fact that more reliable and robust equipment is manufactured today, many project managers will say it is still a sensible approach when selecting plant. Most people will agree that to drive a car at maximum speed for hour after hour on a motorway is not treating the vehicle very well and is inviting trouble. With a concrete pump, if the pressure in the line is so great that the engine begins to labour or falter, pumping should be stopped, otherwise damage may occur.

Wear on pipelines is greater nearer the pump because the pressure is higher the nearer the pipes are to the pump. Worn pipes may still be serviceable and should be placed towards the end of the pipeline where there is less pressure and they can be used safely, depending of course on the degree of wear. Wear must be carefully monitored to ensure that the wear is within safety limits (see section 2.2). Wear is greatest on the outside of bends. These should be reversed from time to time to equalize the wear.

To sum up the matter of wear and how to control it best, it is perhaps a matter of handling equipment carefully and using it sensibly within its recommended limits. This will certainly help to make it last longer.

10

Screed pumping

The mixing and conveying of floor screeding material from the mixing point to the point of placing has in past decades been laborious as well as costly. However, certain types of piston, pneumatic and worm pumps can overcome this problem.

10.1. Conveying floor screeding material

Often the route will be tortuous and access problems will inevitably confront the builder when tackling the finishing processes. Traditional methods of screeding material distribution included the use of hoists, staging or scaffolding with wheelbarrow runs threading their way through already completed areas of the structure. It is impractical to have a hoist positioned to suit every placing position. Inevitably additional costs in temporary works, e.g. wheelbarrow runs with the associated protective measures, are involved. Damage can easily occur when manoeuvring a fully laden wheelbarrow or mortar pram through door openings or narrow passages. These traditional methods obviously involve additional labour costs in distributing the material manually and in the temporary works required.

In addition the safety aspect is important and certain hazards are unavoidable when using such methods. Manually transporting relatively heavy loads by wheelbarrow or any other such method has increased risks when working at heights. A stumble can result in a serious accident.

The rising cost of labour and continuous moves towards mechanization have caused a steady development of plant and equipment in this area of construction work, where indeed it was obviously needed. It was recognized that old methods must be superseded by the new methods. There are various conveying systems in use which facilitate the distribution of the screed material to heights and lengths well in excess of 100 m. Manufacturers claim that they can reach any part of a multi-storey building.

The material is conveyed through small diameter flexible hoses on the most convenient route from mixing station to placing position and the whole operation can be set up quickly, economically and efficiently. Many of the safety hazards already mentioned when using traditional methods are eliminated.

10.2. Types of distribution systems

The distribution systems can be divided into three types: piston pumps, pneumatic conveyors, and worm pumps.

Piston pumps

In the smaller range of piston pumps there are several types which can be termed multi-purpose and even though they are normally classed as concrete pumps, they can be used very effectively for pumping floor screeding material including mortars. Generally they are used in conjunction with a ready-mixed concrete supply. These pumps have higher outputs than the smaller pneumatic units and are able to handle material which has a grain size of up to say 16 mm. For the larger jobs such as very large floor areas where there is scope for high production, Schwing's compact BP750R pump is a typical example of a small piston pump which is suitable for screed pumping as well as concrete pumping

Fig. 114. Piston pump suitable for screed pumping (courtesy F. W. Schwing/Burlington Engineers)

generally (Fig. 114). Another example of this type of pump is the Sidewinder SW30, which is a well proven medium range piston pump used for many applications (Fig. 115). It is trailer mounted with an output of 30 m^3/h and a line pressure of 60 bar. It is powered by an 80 HP four-cylinder water-cooled diesel engine. The Putzkret K 2009, which has an output of 50-60 l/min over 160 m, is a double piston compact unit with the drive block, pump, agitator, hopper and trough mixer lined up in sequence as one unit (Fig. 116).

Fig. 115. Sidewinder SW30 piston pump (courtesy Premier Concrete Pumping Ltd)

Fig. 116. Putzkret K2009 screed pump (courtesy Putzmeister Ltd)

These pumps can also be extremely useful for other applications. They can be used for foam concrete work, slurries and similar applications. Any multi-purpose item of plant on a site is attractive to the owner or user, as it presents opportunities for high utilization.

Mix design for screeds should broadly follow the design for concrete pumping as the mechanics of concrete pumping still apply. Mixes that are too wet, too dry, poorly graded or containing extremes in cement quantities may give problems.

Pneumatic conveyors

Pneumatic conveyors are perhaps more commonly used for screed pumping than any other type and have been developed over the past two decades specifically for this application. The pneumatic conveyor is a simple machine consisting of a pressure mixing vessel and a screw-type

Fig. 117. Typical pneumatic conveyor for screeding material

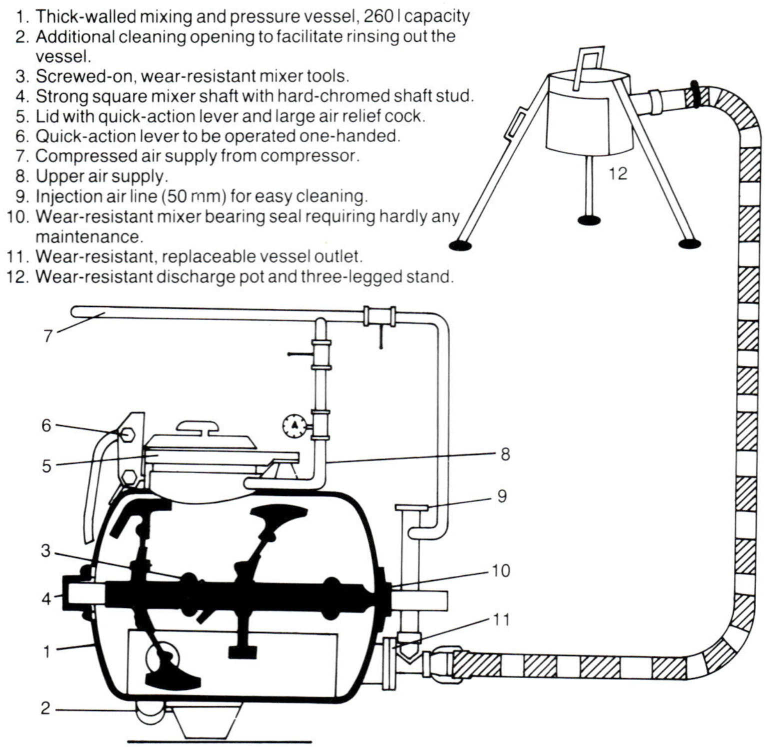

compressor. Sizes of compressor can vary from 3 to 5·6 m^3/min (105–200 ft^3/min), and pressure vessel capacity can be as much as 500 l. The size normally used is about 260 l, which gives an effective mixing capacity of 200 l. The single-shaft mixer with specially arranged blades acts both as mixer and conveyor. The material is loaded into the pressure vessel which is then sealed and compressed air added. The compressed air forces the mix out at the outlet where more compressed air is added from a secondary line to the delivery line. Air cushions are thus formed which transport the mix in an intermittent flow through the delivery line to the placing position. Fig. 117 shows a typical pneumatic conveyor used for screed pumping.

There are various sizes of such units available to suit particular requirements, but a small unit mixing say 2 or 3 m^3/h will supply a sufficient amount of material to keep two people fully employed at the placing point (Fig. 118). The Putzmeister Mixokret (Figs 118 and 119) was developed primarily for conveying floor screeds and Putzmeister claim that it is the most economical solution. Certainly the machine is widely used in the construction industry world-wide.

Fig. 118. Putzmeister Mixokret screed pump (courtesy Pochin Group plc)

Fig. 119. Towing a Mixokret screed pump (courtesy Pochin Group plc)

Putzmeister have made these units since 1966, continually producing a new generation of machine to keep pace with demand. Technical advances in pneumatic conveying have taken place simultanteously with technical development in the construction industry generally and these modern machines are a good example of this.

The Mixokret can be used with or without a scraper unit, but the manufacturers claim increased outputs of up to 30% if these are used. Putzmeister have pneumatic conveyors with compressor ratings of 2–5 m^3/min (70–175 ft^3/min) and mixer vessel sizes of 150, 250 or 500 l. Hydraulically operated skip loaders, manual scrapers, automatic water metering, central lubrication and high pressure cleaning are all available with these machines. Engines are soundproofed to 70 dB. Outputs vary from 2 to 8 m^3/h with pumping heights of up to 30 storeys and distances of 150 m.

If the mixing facility is not required, as would be the case if premixed materials are supplied from a ready-mixed source, the materials can be loaded through the delivery line with compressed air. Setting up time for the Mixokret is minimal at 15–20 minutes. It is capable of supplying various different working points even on large sites, from one position. With regard to the use of skip loaders, checks made under working conditions have proved that this can increase production significantly, provided that a relatively dry sand is available for the mix.

Other manufacturers produce various types of pneumatic pumps for this application, for example, the Transmat screed pump/mixer (Fig. 120). This machine is designed on the same principle as most other pneumatic conveyors and is

Fig. 120. Transmat screed pump/mixer (courtesy Sidewinder Concrete Pumps Ltd)

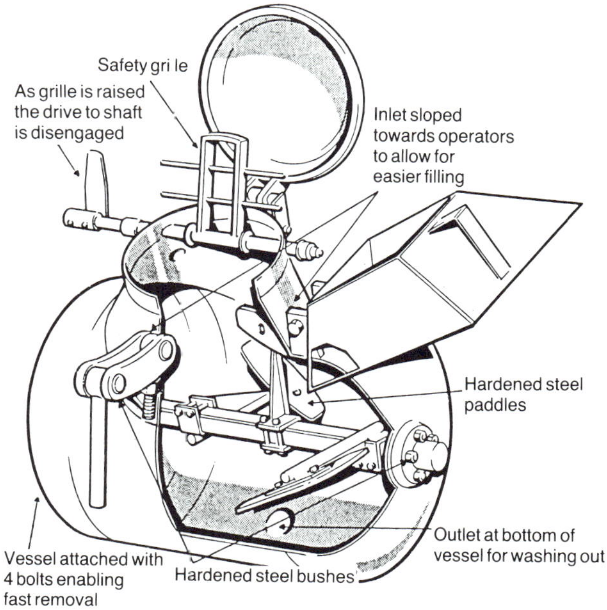

Fig. 121. GSP mixing chamber (courtesy Greenhill Plant)

suitable for mixing and conveying dry or semi-dry sand/cement screeds up to a maximum aggregate size of 15 mm. It has an output of 3 m^3/h.

The only screed pump currently made in the UK is the GSP pneumatic pump, which has been developed and marketed by Greenhill Plant, a company based at Twickenham, Middlesex. The design of the mixing vessel, which is shown in Fig. 121, facilitates cleaning out and reduces the time normally required for this work. The capacity of the mixing vessel is 281 l and the vessel can be detached readily by removing four bolts and one air union. A safety feature is that the safety grille can only be raised if the drive to the mixing vessel is in neutral. The company has produced this equipment with safety very much in mind, something which is to be commended.

Two types of air compressor can be used for this compact but powerful screed pump: the Compair Zitair 175 or the Atlas Copco XA85; both have a capacity of about 5 m^3/min (175 ft^3/min). The 72 BHP engine (either Perkins

Fig. 122. GSP pneumatic screed pump (courtesy Greenhill Plant)

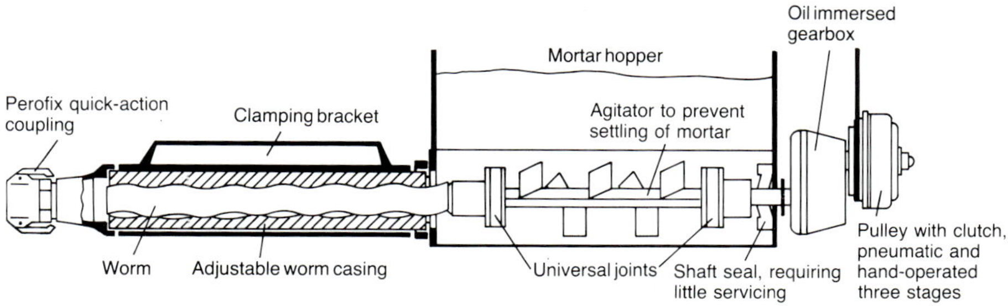

Fig. 123. Mortar worm pump (courtesy Putzmeister Ltd)

or Deutz) has 30 HP surplus available to deal with the mixing vessel requirements. The manufacturers claim that this unit will convey screed material over a distance of 160 m and that it will pump even difficult semi-screed/concrete mixes (Fig. 122).

Worm pumps

Worm pumps are used extensively for conveying and spraying the less difficult mortars, premixed mortars and various plasters. They are simple yet effective machines which are easy to clean and maintain. There are very few parts which can wear out and costs in this respect are very low provided that the pump is not worked at high pressures. If the working pressure is kept within the bounds of 10–15 bar, wear is minimized. Outputs from this type of machine vary from 3 to 10 m^3/h. Figure 123 illustrates a typical worm pump.

Most of these pumps have mixer attachments and variable output control facilities. The pump is driven by either a diesel or electric motor. There are both stationary and trailer mounted models. The advantage of worm pumps is that they are surge free and convey the material smoothly, which is extremely useful for certain applications, e.g. lightweight concrete and sprayed concrete using 0–8 mm aggregates.

Putzmeister's P11 worm pump (see Fig. 24) has an adjustable stator and it is claimed that this substantially reduces wear on this integral part of the pump. This unit may be equipped with a 170 l capacity mixer, a 200 l skip and has an output of 3 m^3/h — the output normally required on the smaller jobs. It operates at low pressures, around 10 bar, and will convey material to heights in excess of ten storeys, given a suitable mix design. Putzmeister also produce another version of the worm pump. This has a higher output of up to 20 m^3/h (Fig. 124).

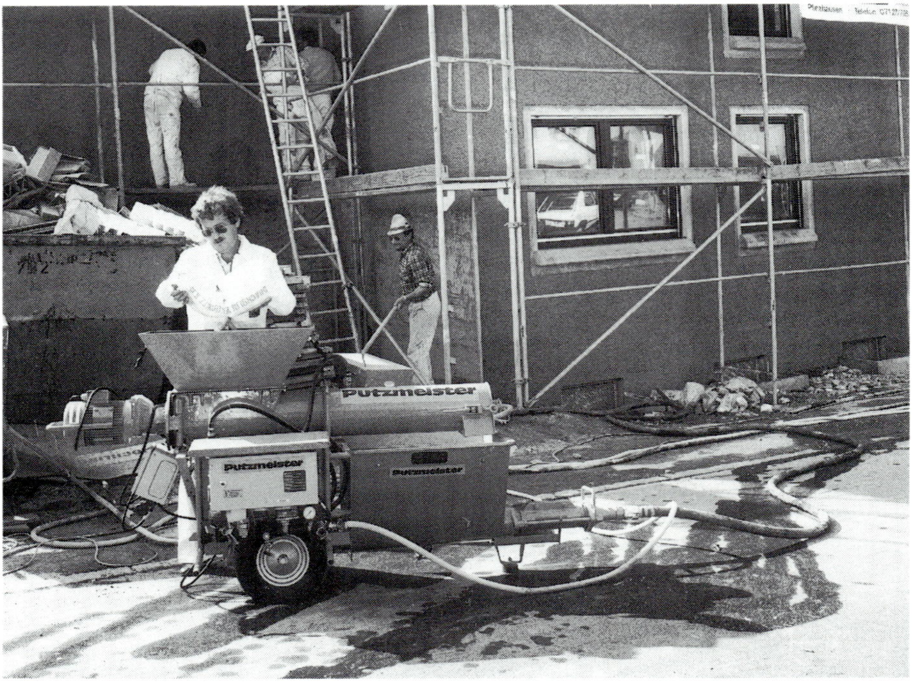

Fig. 124. S8 mortar/worm pump (courtesy Putzmeister Ltd)

10.3. Screed mixes

A typical floor screed mix would include sand with cement, anhydrite or gypsum bonding agent. Aggregate grading curves recommended by Putzmeister for use with the Mixokret are shown in Fig. 125.

Lean mixes are easier to convey than rich mixes when using pneumatic conveyors. Floor screed mixes are usually fairly stiff (bearing in mind that an intermittent flow is required in the delivery line) which will provide material of the correct firm consistency at the placing position.

Type of sand is important. It should be sharp, washed and must not contain any impurities. Screed mixes of ultra dry or fine sand are most likely to be unpumpable. Well graded sands with aggregates up to 8 mm are suitable, with only a small amount below 0·15 mm (Fig. 125).

The cement/sand ratio is invariably about 0·25 and the water/cement ratio is between 0·35 and 0·5 depending on the fines content. Plasticizers may be added as required. For an anhydrite floor screed, having the correct sharp sand is essential and the cement/sand ratio should be nearer 0·33 in

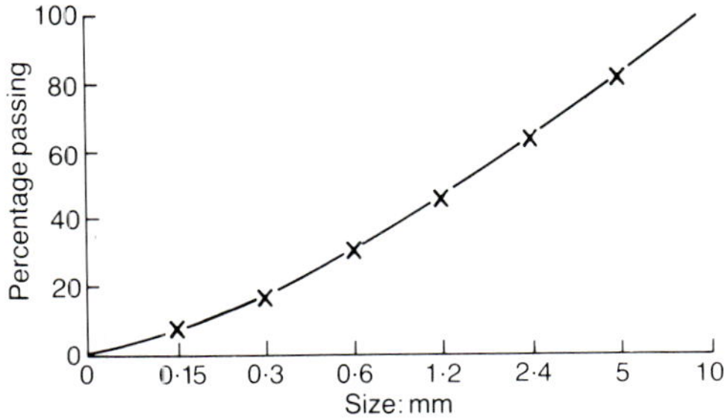

Fig. 125. Typical grading curve for use with the Mixokret (courtesy of Putzmeister Ltd)

this case. The vclume ratio of 0·25 is equivalent to one bag of cement per mixing vessel load, which is a convenient measure.

10.4. Safety aspects and procedures for pneumatic conveyors

The use of pneumatic pumps of any type requires formal operator training and most responsible firms ensure that proper written procedures are drawn up and that the operator is fully trained. Bearing in mind that compressed air and dangerous moving parts (e.g. the mixer assembly) are involved, certain precautions are necessary. First, the mixer grille must be firmly fixed in position before the machine is operated and nothing must be put into the mixer until the engine has been stopped and the agitator shaft has ceased rotating. Second, in relation to the use of compressed air, all pressure must be released before any part of the machinery or delivery line is uncoupled. These two simple precautions should, if heeded, significantly reduce the number of accidents.

10.5. Hints for the pneumatic conveyor operator

Blockages

If the pump is pumping normally, the vessel pressure gauge will read 4–6 bar. If the pressure gauge is showing 6–7 bar the indication is that there is a blockage, usually referred to as a high pressure blockage. If the pressure is zero, there is likely to be a blockage at the outlet end of the mixing vessel.

To clear a high pressure blockage, leave the air pressure on and check the pipeline by hand. The pipe will normally feel hard because of the air pressure. When a soft area is located, the blockage will be at the junction of the hard and soft areas. On no account must the pipeline be opened until the air valves have been switched off, the vessel lid air release valve opened, all the air pressure released and the vessel lid opened. It is now safe to open the pipeline near the blockage, which can now be removed.

When dealing with a low pressure blockage, the valve to the air delivery downpipe should be closed and pressure should then build up in the vessel. This may clear the blockage. Otherwise, the pressure gauge will rise to about 6 bar. In this event, close both air valves, disconnect the outlet hose from the vessel and clear the vessel outlet.

Cleaning out

Pumping of the final mix should continue longer than normal to clear the mixing vessel as much as possible. The vessel is then opened following the set procedure to clear a high pressure blockage, and the vessel filled with water. The lid is now closed and the machine run for at least five minutes. The water is then flushed through the pipeline by operating the main air delivery valve to the vessel; the bottom air line is closed. Any water sediment in the vessel can be discharged by removing the vessel drain cap.

To clean out hoses, two wet sponge balls are inserted into the hose and carefully forced through using the main air supply in the same manner as for flushing water through a pipeline.

Possible problems

Failure to convey material, indicated by low pressure, may be due to a blockage at the vessel outlet. The type of sand may be unsuitable, or possibly too many fine fines in the mix.

Insufficient mixing can cause a blockage. This may be due to misaligned or broken agitator blades.

An air loss in the system can also be the cause of a blockage. Air losses can be due to faulty (worn) hose joints or a faulty lid seal.

Failure to convey material, indicated by high pressure, can be caused by a blockage in the vertical down pipe, inhibiting the forming of air cushions in the delivery line. The mix may be too dry or the line blocked by a foreign object. The tripod outlet is also a point where there might possibly be a blockage. It should be noted that the appropriate air valve must be closed before inspecting the down pipe to the vessel outlet.

10.6. Other applications of pneumatic conveyors

Although pneumatic conveyors such as the Mixokret primarily convey floor screeding material, they can also convey sand, gravel, fine aggregate concrete and even earth. If concrete is to be pumped, care must be taken to ensure that the correct hose diameter is chosen for the corresponding aggregate size. It is also necessary to use a larger capacity compressor for conveying concrete.

It has been said that one person using a bucket could move only 3 m^3 of sand in a working day from ground level to second floor level. One person with a Mixokret machine can move 12 m^3 under the same circumstances. The sand must be clean. Wet, sticky pit sand is unsuitable. The advantages of being able to convey pneumatically relatively large quantities of bedding sand for floor tilers who may be working many storeys up a building are obvious. It is clear that there must also be significant cost savings and other advantages.

Conveying gravel and fine aggregate concrete is also possible with this type of machine, although larger aggregates can be used with the 500 l units. The performance will not be as good as a conventional concrete pump, but in certain circumstances pneumatic conveyors can be useful for small scale, low output requirements which are coincident with the need to convey screed.

Earth (humus) was conveyed in the Berlin Zoo through 65 mm hoses over a distance of 80 m, and the output was 5 m^3/h. Although there may not be a large market for such an application, this does illustrate the versatility of these machines.

10.7. General points

The installation of a small mixing/pumping station for the floor screed supply to a contract offers some scope for ingenuity. Careful planning and consideration of the most

Fig. 126. Overhead loading arrangement for screed pumping

economical routes for delivery of the material will pay dividends. If there are constant daily requirements for material, there are benefits to be obtained from providing mechanical assistance in loading from an overhead hopper which is filled by a conveyor (Fig. 126).

Building work has its well-known problems with finishings and such statements as 'it's the finishings that take the time' are commonplace. Any means of speeding up this work while maintaining quality should be considered. Moving the floor screed material quickly and efficiently to all parts of the structure is therefore important. It is worthwhile for the contractor to compare the cost of operating a screed pump with the cost of traditional methods which incur their own additional direct and intangible costs, as previously described.

However, the contractor first of all must make sure that the most appropriate machine will be used. The significant range of screed pumping units which are available as small piston pumps, pneumatic conveyors, or worm pumps, offers a wide choice for the user. Choosing the machine to suit the specific requirements or circumstances (e.g. high, medium or low volume output, height and length involved in conveying material, type of aggregate and mix specification), is an important step. It is sensible for the contractor to consult with the manufacturers or pump hire companies. Following this, a true cost assessment (see chapter 6) can be carried out with all the cost factors taken into consideration. The machine should be the least of the worries: two decades of continuing development and improvement have guaranteed the reliability of the modern screed pumping plant and equipment.

Whether using pneumatic or piston type screed pumps, conveying screeding material need no longer pose a problem for the contractor and indeed most of them now take advantage of this method. It is widely recognized that moving the screed quickly and in good condition to the exact placing point is the sensible approach. It also makes sense commercially.

11

Sprayed concrete

Sprayed concrete originated in the USA at the beginning of this century when Dr Carl Akeley, a famous hunter and naturalist, devised a method of constructing plaster models of large animals. He sprayed plaster on to the wire framework of his models with a special 'gun', using compressed air. The method was to contain the material in the gun, use compressed air to force it out of the nozzle and wet it with the correct amount of water as it was blown on to the framework. He was able to form a strong thin coating of plaster which would not drop off the frame. There was no longer the problem of wet plaster setting before it was ready to be placed, because the gun was filled from a heap of premixed dry material.

Thus the construction industry was later to benefit from the ingenuity of someone who was not even remotely connected to the industry; the equipment designed by Akeley was successfully used to patch up damaged concrete using a sand/cement mix.

In 1911 a patent was issued for the equipment, which was called the Cement Gun, and for the material, which was called Gunite. In 1915, in Allentown, PA, USA, the Cement Gun Company was formed to build the machines and to carry out the Gunite process. The Allentown Pneumatic Gun, Inc., is the successor to the original company. Until 1971 its registered trademark was Gunite, but the name has now come into common usage in the construction industry.

11.1. Sprayed concrete processes

In the USA, sprayed concrete is usually referred to as shotcrete. The ACI has taken the lead in the subject by producing specifications and technical papers. The ACI's definition of shotcrete is 'mortar or concrete conveyed through a hose and pneumatically projected at high velocity on to a surface.' In the UK it is usual to refer to the process as gunite if the maximum aggregate size is less than 10 mm. If the aggregate size is greater than 10 mm the process is called shotcrete. Ryan[63] provides much useful and interesting information on applications and equipment as well as practical considerations of nozzle techniques, operator duties and safety precautions.

Methods of spraying concrete

There are two methods used in sprayed concrete: the dry process and the wet process. Up until the 1950s only the dry process was used and it was referred to as guniting, and is still called guniting. The wet process was developed in the 1950s. As the names imply, in the dry process the dry materials are conveyed to the gun nozzle, whereas in the wet process the already wet (hydrated) materials are pumped to the nozzle.

Dry process

Figure 127 shows the dry process or dry mix method. The material is batched and mixed without the addition of water. It is placed in the so-called pneumatic gun where it is pressurized and an even flow of the material is fed into a high velocity compressed air stream and conveyed through flexible hoses to the nozzle. At the nozzle sufficient water is added to hydrate the mix and provide the desired consistency for the projection of a stream of material at high velocity on to the surface to be treated, usually referred to as the substrate. As no water is required to enable the material to be transported to the gun, the process involves low water/cement ratios with the result that it is very suitable for placing limited thicknesses on to vertical or overhead surfaces. Care must be taken in adding the correct amount of water as the correct range of water/cement ratios is limited. Too much water will increase the workability so that the material will slump, whereas too little will inhibit

Fig. 127. Dry process

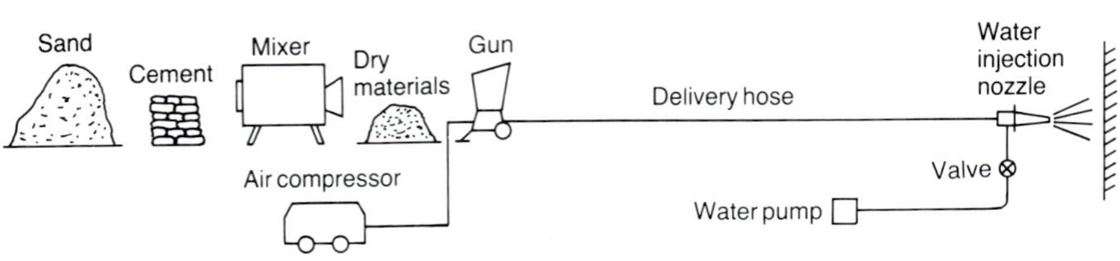

compaction into a homogeneous mass. Normally aggregate sizes not exceeding 10 mm are used in this process, although it is possible to use larger sizes up to 20 mm.

Aggregate/cement ratios vary between 3·5 and 4 by weight. The placed material will be richer in cement than batched concrete as the unavoidable rebound is mainly aggregate. Performance characteristics include good density and high strengths which are normally between 40 and 50 N/mm^2. The material bonds excellently to well prepared surfaces.

Wet process

Figure 128 shows the wet process. The main difference between this method and the dry method is, of course, that the material is already hydrated when it reaches the nozzle. Water is added to the mix before it is pumped to the nozzle and the compressed air for compaction is introduced at the nozzle. The concrete or grout for this process is usually supplied from ready-mixed concrete trucks, although the concrete or grout is sometimes mixed on site.

With this process it is possible to have water/cement ratios as low as 0·4, because plasticizers can be used. As the material is being pumped, specifications should be generally as those required for a pumpable mix. However, it is usual to have high cement contents between 350 and 450 kg/m^3, which gives 28 day cube strengths of between 35 and 55 N/mm^2. Figs 129 and 130 show the ACI recommendations on grading envelopes for sands and aggregates[64] which followed a symposium of 13 papers forming a guide to American practice.

Henderson[65] devotes a section of his book to sprayed concrete and states that admixtures such as microsilica and accelerators can be added to the mix if required. In the wet process the admixture will be introduced in the usual way, normally as a liquid. In the dry process it can be added as a powder into the dry premix, as a liquid with the water at the

Fig. 128. Wet process

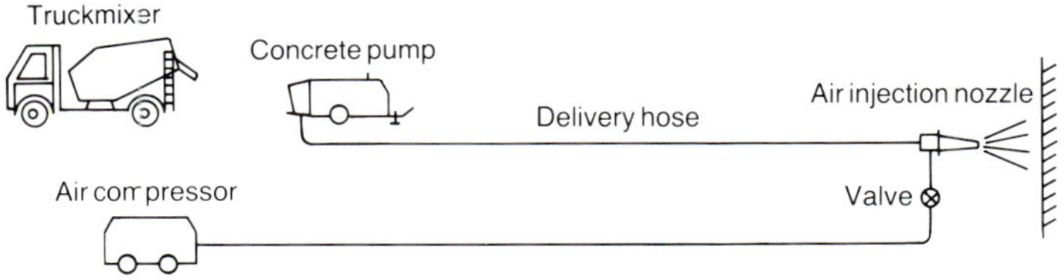

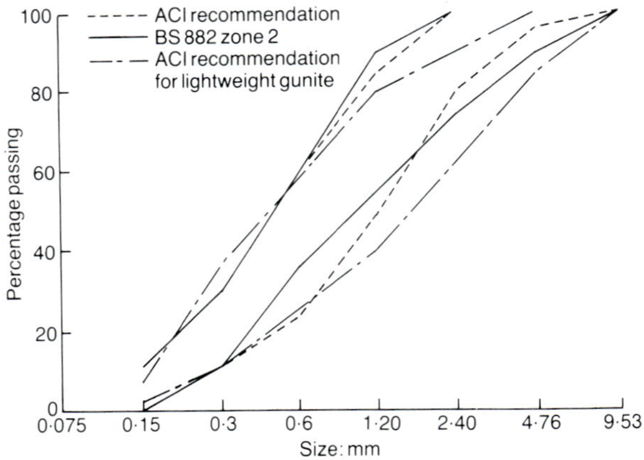

Fig. 129. Grading envelopes for gunite sands

Fig. 130. Grading envelopes for aggregates up to 20 mm

nozzle or as a separate injection at the same point. When used carefully, admixtures can produce successful results in sprayed concrete work. However, some admixtures which have been used advantageously in normal concrete may not have the same effect when sprayed.

Wet sprayed concrete is a structural material and is applied in layers, sometimes built up over several days if necessary, to thicknesses of over 200 mm. Bonding between layers is excellent if the underlayer is left as sprayed. Trowelling should be carried out with caution, as anything other than a very light application could cause surface cracking and density loss. It is possible, however, to achieve

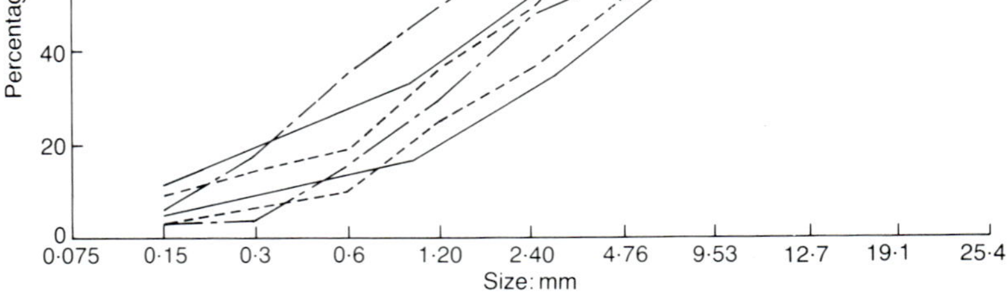

tolerances of ± 10 mm over 3 m lengths of plain flat surface. More complicated surfaces require special provision.

Other differences between the two methods

Although the dry process is used more commonly than the wet process, the market share of the latter is steadily increasing. Each system has its advantages and disadvantages, but in the wet process problems from rebound and dust are reduced. In dry weather, dust can be a problem when using the dry process.

It is sometimes argued that it is not a good thing that the control of the mixing is left to the operator in the dry process, as the operator is adding the water as the spraying continues. Most firms, however, ensure that their operators are properly trained. With experience, operators are well able to judge the amount of water required. It is quite apparent, when operating a spray gun and closely observing the material as it is applied to the surface under treatment, if there is too much or too little water being added. On the other hand, the wet process operator has simply to apply air at the nozzle, as the water has already been added before the concrete reaches the pump hopper.

The wet process has higher outputs — up to 10 m^3/h — than the dry process and is therefore more suitable for larger volume jobs. It is normally specified for thicknesses in excess of 50 mm. For thin applications and for minor repairs, however, the dry process appears to be better.

As the dry process material is dry until actually placed it may be used intermittently, i.e. in a situation where it may not be possible to spray continuously. A truck load of ready-mixed concrete which has been supplied for the wet process cannot be left too long and continuous working is essential.

The choice of whether to use the wet or dry process will depend on parameters such as type of job, specification and accessibility. The customer is advised to consult the specialist with clear specification requirements of all aspects of the work. If this is done, the customer has every chance of completing the job satisfactorily with the more appropriate system. However, most types of work can be carried out using either process. There is no clear-cut differentiation and both systems can be used on any type of job, with the possible exception that high output jobs are more likely to be carried out using the wet process.

11.2. Why specify sprayed concrete?

The sprayed concrete companies will claim that the properties of sprayed concrete are better than those of concrete placed by other conventional means in certain circumstances. If unusual shapes are required, there is a significant advantage in using sprayed concrete because almost any shape can be formed without using formwork. The method of placing — depositing the concrete material with some force on to the placing area — produces dense, homogeneous structures which are particularly suited to the normally relatively thin linings formed by the process.

Use of formwork

It is well known that formwork costs are a high percentage of the total cost of a concrete pour. The costs of a pour can be divided into four main areas: formwork, steel-fixing, concrete materials and placing costs. Individual percentages will vary depending on the shape and thickness of the pour, quantity and degree of steel-fixing difficulty, access etc. All these items have to be considered and if formwork can be reduced or even eliminated, costs can be reduced. For example, sprayed concrete could be used for a swimming pool which has curved walls and floors.

Illingworth[66] gives costs of formwork, reinforcement and concrete as percentages of the total concreting operation. Although these figures do not specifically concern sprayed concrete applications, they nevertheless illustrate the point that formwork on average accounts for the major share of the total cost of the concreting operation (Table 29). However, these figures must not be used as a general rule and refer to specific situations recorded.

The ability to cope with the most intricate of shapes without the use of expensive formwork must appeal to the designer. Also the ability of the system to facilitate a variety of finishes is an added advantage.

Table 29. Cost breakdown for a reinforced concrete frame of a typical six-storey office building; figures include total cost of materials and are expressed as a percentage of the total cost

Component	Cost				
	Average	Columns	Beams	Slabs	Walls
Formwork	39·5	24·4	33·0	47·7	53·1
Reinforcement	33	60·9	56·3	7·1	23·8
Concrete	27·5	14·7	10·7	45·2	23·1

11.3. Finished concrete qualities

Provided that the substrate surface has been prepared properly to receive the sprayed concrete, the material forms an excellent bond and the method ensures that voids are filled satisfactorily. High strength concrete with high impermeability is claimed and has been proved in many cases. The mix is already rich in cement before application, and is then further enriched due to the inevitable rebound, which is mostly aggregates. The low water/cement ratios used increase the strength.

Increasing the thickness by careful application of several layers minimizes thermal stresses in such sections. Bonding between layers can be extremely effective if carried out correctly.

There is no doubt that the finished concrete will be good quality, high strength and homogeneous, provided that the mix is properly designed and the material applied by an experienced operator. This has been demonstrated many times by cores taken from test panels set up under the same conditions as a sprayed concrete operation.[63] It is not usual to use cube testing in sprayed concrete work.

11.4. Access problems minimized

There are numerous examples of how the use of sprayed concrete equipment can ease access problems, particularly on the small repair jobs where access is difficult with other plant. One person equipped with a spray gun can gain access to the most difficult of places. Yet for major works such as tunnel linings or tunnel repairs, track mounted or wheeled shotcrete mobile units, which include the concrete pump, are able to operate inside tunnels if the tunnel is large enough. Booms have a reach of over 10 m. Apart from tunnel work, sprayed concrete is also used extensively on many other applications such as canals and retaining walls.

For example, the Putzmeister Buffalo shotcrete machine, which is a wet process machine (Fig. 131), is controlled by the operator from a well protected cabin close to the placing area and there is excellent visual and operating control. The cabin and shotcrete arm can rotate through 360°. Such machines are used extensively in new tunnel construction as well as repair work. For other jobs presenting access problems, both wet and dry systems using conventional sprayed concrete equipment are suitable. Where it is not possible to deliver materials either by truckmixer or for

Fig. 131. Wheel-mounted Buffalo shotcrete unit (courtesy Putzmeister Ltd)

batching close to the placement point, inaccessible positions can be reached by the use of delivery lines in excess of 300 m horizontally and 150 m vertically. A well laid out static line can sometimes be the right answer.

11.5. Sprayed concrete plant and equipment

Types of machines and manufacturers

There are various types and sizes of sprayed concrete units available throughout the world, for use with either the wet or dry process. Other units may be used with both systems. There is a wide choice of well developed and reliable units, ranging from small dry process machines with maximum outputs of just over 3 m^3/h to the larger machines (both processes) of up to 12 m^3/h and capable of using 25 mm aggregates.

In the UK, the Aliva Company, based at Hatfield, Herts, produces a range of spraying machines for both processes. The 240, 250 and 260 dry process machines have outputs ranging from 3·4 to 9 m^3/h and are capable of conveying material up to 300 m horizontally and 100 m vertically.

Aliva manufactures a multi-purpose machine, the Aliva

Fig. 132. Type 280 Universal unit, for both wet and dry processes (courtesy Aliva Ltd)

280 Universal (Fig. 132), which can be used for both the wet and dry processes. The changeover takes only a few minutes with no conversion or extra accessories being required. The machine has an adjustable hydraulic drive which can set outputs of between 2 and 11 m^3/h for either process. This company also produces and distributes hydraulic spraying arms. These can be wheel mounted (Fig. 133) and are particularly useful in tunnel applications.

Fig. 133. Type 304 robot spraying arm (courtesy Aliva Ltd)

Fig. 134. Topshot wet shotcreting unit (courtesy F. W. Schwing/Burlington Engineers)

Both Schwing and Putzmeister produce concrete pumps and equipment for sprayed concrete, mainly for the wet process. Schwing's multi-purpose trailer concrete pump, the BP750R, is used extensively with shotcreting equipment. The firm has developed its Top Shot wet concreting system (Fig. 134) in conjunction with this pump. The system includes an admixture tank mounted on the pump chasis; the dosing pump is integrated into the hydraulic system. Schwing claims that rebound is minimized with this unit, even on overhead applications.

The system can be truck mounted; an aerial basket is fitted to a telescopic arm which the manufacturer says is constructed on the principle of the human arm and hand. The arrangement is hydraulically operated by remote control, is very manoeuvrable and can be controlled easily. Putzmeister's BSA 1002D trailer pump, which has a theoretical output of 20 m^3/hr, has been used with considerable success in wet shotcreting. A major part of the new canoe slalom course at the UK National Water Sports Centre, Holme Pierrepont, near Nottingham, was constructed in 1986 by Shotcrete Services using a Putzmeister 1000D pump (Figs 135 and 136). Differing requirements of curved concrete forms necessary for such work meant that conventional formwork

was neither practical nor an economic proposition and 1000 m^3 of sprayed concrete provided the solution. Putzmeister also manufacture the shotcrete Buffalo range of machines for sprayed concrete, which has already been described.

Other manufacturers include the Allentown Pneumatic Gun, Inc., which was mentioned earlier. The Allentown wet and dry units are very similar to those already mentioned. The company produces one and two axle dry process

Fig. 135. Canoe slalom course at Pierrepont, Notts, under construction (courtesy Shotcrete Services Ltd (now Sprayed Concrete Ltd))

Fig. 136. After completion of shotcreting at Pierrepont, Notts (courtesy Shotcrete Services Ltd (now Sprayed Concrete Ltd))

Fig. 137. R-900 gunning rig with N-2 gun (courtesy Allentown Pneumatic Gun Inc.)

machines which will produce 10–12 m^3/h of dry mix (Fig. 137). In place, this will give 6–8 m^3/h, allowing for compaction and rebound. For this maximum output, Allentown states that up to 60 mm dia. hose and a 25 m^3/min (883 ft^3/min) compressor are required. Normally the machines operate at outputs of up to 5 m^3/h using smaller diameter

Fig. 138. Model 40 Powercreter (courtesy Allentown Pneumatic Gun Inc.)

hoses and a compressor in the range 12·7–17 m^3/min (450–600 ft^3/min).

The Allentown Powercreter pump (Fig. 138) is used in the wet process and this has outputs of 25–40 m^3/h. The capacity of the air compressor need only be 4·2–7 m^3/min (150–250 ft^3/min). It is claimed that steady production rates of between 15 and 20 m^3/h in place in favourable conditions can be achieved on work such as slopes and walls (as against overhead work), and large projects can be carried out economically. For overhead applications Allentown recommends an accelerator admixture at the nozzle to provide quick setting.

The Italian manufacturer, CIFA SpA, includes shotcreting equipment in its range of concrete pumping units. The portable shotcrete pump model PCS209 (Fig. 139) has an output of up to 30 m^3/h and can operate at a maximum concrete pressure of 93 bar. The company also produces a lorry mounted shotcrete unit with a special two or three-section boom and concrete pump, called the Spritz system (Fig. 140). The turret slewing range is 360° and boom working lengths are up to 10 m. Two-section boom attach-

Fig. 139. PCS209 portable concrete pump shotcreting an embankment (courtesy CIFA SpA)

Fig. 140. Spritz shotcrete unit with three-section boom (courtesy CIFA SpA)

ments for shotcreting are also available from CIFA for fitting to other construction equipment such as forklifts, loading shovels, mobile cranes, etc. The pumping unit can be supplied separately or as part of a complete package.

11.6. Applications for sprayed concrete

Sprayed concrete applications may be considered as coming under five headings: strengthening and repair work; water retaining structures; underground linings; protective coatings and linings; and new construction.

Strengthening and repair is a natural application for sprayed concrete and much of the world-wide market is concerned with remedial work to damaged concrete. The damage could be caused by the weather, action of the sea, subsidence, fire or merely wear and tear. Consequently this application would include such work as repairs to bridges, arches and cooling towers where finishes to remedial work are required. Buildings also come under this heading, together with underpinning such structures. Coastal erosion

Fig. 141. Formation of a lake in a private garden (courtesy Windsor Concrete Pumping Ltd)

and damage due to wave and tidal action is a continuing problem and sprayed concrete has an important part to play in the repair work. Injection grouting coastal defences is generally followed up by sprayed concrete rendering to give the surface a smooth finish.

Water retaining structures, which include swimming pools, concrete tanks and water towers, canal and drainage channel linings, invariably make use of sprayed concrete. The construction work in the freeform design which is now a significant part of pleasure park water retaining structures, has been facilitated and made more economical because of the higher availability of sprayed concrete. Shuttering costs to create the same shapes would be prohibitive.

The construction of lakes and swimming pools with curved walls and floors (Figs 141–143) is an important market for the sprayed concrete industry as well as major projects such as reservoirs, aqueducts and sea walls. These structures require configurations more often than not which will rule out normal shuttering for purely cost reasons.

Underground linings will most certainly include tunnels of all types, storage caverns and basements, underground roads and large diameter sewers. The development of the large

Fig. 142 (facing page). Swimming pool under construction using wet process (courtesy Windsor Concrete Pumping Ltd)

Fig. 143 (facing page). Completed swimming pool constructed using wet process (courtesy Windsor Concrete Pumping Ltd)

Fig. 144. Embankment construction using wet process (courtesy CIFA SpA)

shotcrete mobile units, as already described, for use mainly in tunnels has proved that there is a considerable market for this type of work in all parts of the world. Although the wet process is used mainly in tunnel works, the dry process still retains a high share of this part of the market and many underground structures in need of repair and rendering will be treated successfully using this system.

Fire protection coatings using spraying techniques are widely used. Other applications in the category of protective coatings and linings include corrosion protection to steelwork and pipework, bunker silo and refractory linings. Other forms of protection include coatings or layers of concrete to retaining walls, embankments and dams. Sprayed concrete will effectively deal with the small unstable slope in a private garden or with the stabilization of a large railway or motorway embankment. Such work may often entail the use of drainage channels built into the slopes for excess water to run off (Fig. 144)

The sprayed concrete firms will always hasten to point out that much of their work is in fact new construction and not simply repairing old structures. Certainly there is evidence to support this in the number of new structures in which sprayed concrete has been used in at least part of the construction. Shell roofs and domes can be readily formed by this method, as only the underside formwork is required.

In civil engineering foundations, bored piling retaining walls or secant piling walls are invariably faced with concrete, for which sprayed concrete is well suited. Sprayed concrete is also well suited to the facing work required for diaphragm walls. The dense rendering is ideal for dealing effectively with any water seepage between piles or at construction joints in the walls.

Sprayed concrete is playing an effective part in many other types of new construction such as swimming pools, sea walls and retaining walls where complex shapes are required. Considerable use of the system is also being made in both private and industrial building work. But it is perhaps in the freeform design field that sprayed concrete can be shown to its best advantage (Fig. 145).

11.7. Fibre-reinforced sprayed concrete

Steel fibre reinforced concrete (SFRC) is now widely used and is applied by spraying. Glass and polypropylene fibres are also used, but to a far lesser extent than steel fibres. For certain applications there are considerable advantages in incorporating these materials into the mix, both in respect of the quality of the concrete in its hardened state and in the savings which can be made from reduction in the materials in the permanent works, i.e. volume of concrete, reinforcing steel bars or mesh. In addition, the temporary works such as staging or scaffolding required to place reinforcing steel in, say, a tunnel roof or walls, can sometimes be completely dispensed with, thus affording significant savings in labour and hire of equipment. These aspects will be dealt with in more detail later.

The use of fibre reinforcement in sprayed concrete was developed in the late 1960s. Steel fibres are most commonly

Fig. 145. Freeform design in gunite and Styrofoam of new medical clinic, USA (courtesy Allentown Pneumatic Gun Inc.)

used and it is estimated that the annual world-wide usage is in excess of 100 000 t.

Fine steel fibres are added to the concrete mix and the fibrous mass is then fed to the gun. Either the wet or dry process can be used. The fibres are added by random dispersion or by using special metering equipment. Fibre dosage is usually defined as percentage by weight or as mass per unit volume: the former is more generally used. Quantities in the range 3–5% by weight are usual, but mixes containing higher quantities can be successfully sprayed.

Fibre types and sizes vary considerably. Steel fibres can be cold drawn, slit sheet, or melt extracted. The latter are easier to apply because they are more ductile than the cold drawn types which can cause blockages by balling. The latter can also create excess wear to the equipment. The sizes of fibres vary from 12 to 50 mm, but normally a size of about 25 mm is used. Shorter fibres are easier to mix and spray, but the properties of the concrete are not as good.

Slenderness of the fibre is defined by its aspect ratio (AR) which is the fibre length/diameter ratio. For the normal 25 mm long fibre, the diameter should be within the range 0·4–0·5 mm. When surface fibre corrosion has to be avoided, say in the case of most marine applications, low cost stainless steel fibres are obtainable.

Advantages of using steel fibres

The SCA claim[67] that the quality of the finished concrete product is enhanced in several aspects: the relatively thin layers of sprayed concrete have increased flexural strength, are tougher and have more energy absorption qualities. Thus the concrete has better resistance to impact, abrasion and erosion. Resistance to cracking also has particular benefits for high temperature applications such as thermal linings or fire prevention protective coatings. A flexural strength range of 5–7 MN/m^2 is readily obtainable, with much higher figures quoted as being possible, depending on the type of fibre, shape and dosage. Compressive strength increases over normal unreinforced concrete are also claimed, which are typically up to as much as 50%.

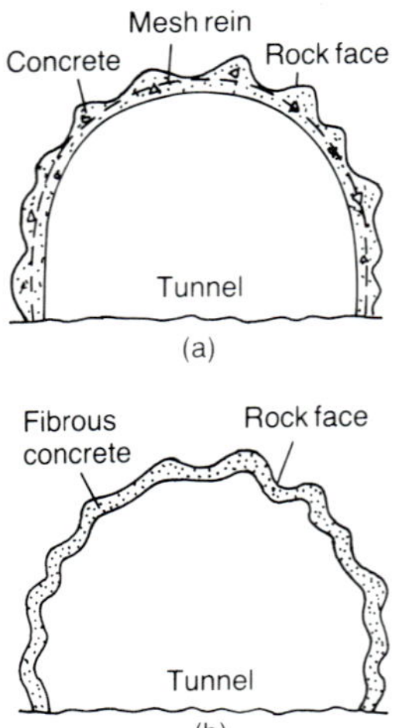

Fig. 146. Materials savings using fibrous concrete: (a) mesh reinforcement sprayed lining; (b) steel fibre sprayed lining (courtesy Sprayed Concrete Association)

The reduced cost of materials, e.g. in concrete and reinforcing sheet, can be substantial. Where normal reinforced concrete is used, a thinner layer of fibre mix can maintain the same strength. The SCA[67] claim that the reduction in thickness can be up to 50% in some instances. In larger projects such as mining and tunnelling, overall contract time savings can be significant.

If reinforcing steel is required for, say, a tunnel lining, the cost of providing and erecting the staging and/or scaffolding required for steel-fixing should be considered. The cost of temporary works in building and civil engineering can be very considerable, as described by Illingworth.[66] Any saving in this respect must always be given consideration and a suitable mix design containing steel fibres might not only eliminate much of the temporary works costs but also offer savings in the layer thickness.

Fixing mesh reinforcement to an irregular rock surface is not easy and large voids may be left which have to be filled. A further advantage is in time savings, which could be very substantial when related to plant and equipment hire, labour and overheads costs (Fig. 146).

Applications for SFRC

SFRC has been used in most of the applications which have been described already where conventional unreinforced material is used. Its advantages, however, lie principally in tunnelling or associated projects where costs can be saved and in thin layer applications such as domed structures, deteriorated concrete surface repairs and corrosion protection to structural steelwork. Marine works where stainless steel fibres are recommended, is also an important area of construction work where SFRC is particularly effective. If steel fibres are used in the mix, the improvement in thermal shock resistance makes this type of mix more suitable for high temperature linings.

11.8. Technical information on sprayed concrete

In the USA, the ACI maintain and publish detailed specifications on materials and procedures used for sprayed concrete.[68,69] In the UK, the SCA, which has been in existence since 1976, has done much to promote the technique and has produced guidelines and specifications for both the wet and the dry processes. Members of the SCA have actively assisted the CS in the preparation of a number of publications including guidelines and specifications for both systems.[70]

The SCA's published information, guidelines and specifications are excellent, as they have been drawn up from the experience gained by its members. The SCA's active technical committee ensure that it keeps pace with technological advancement, both with regard to quality control and plant and equipment.

11.9. Operating procedures and hints on sprayed concrete

As both wet and dry processes rely heavily on the expertise of the person handling the nozzle, it is imperative that the nozzle handler is not only fully trained and experienced in operating the equipment, but understands the principles of the mix design and is aware of the relevant codes of practice. As with any job in the construction industry there is a right and wrong way of doing it. Adhering to the correct operating procedures will enable the nozzle handler to carry out the work speedily and efficiently and to produce a first class job to the correct specification. Most sprayed concrete firms will have their own operating procedures. These might well follow the SCA's guidelines and specifications, which provide a wealth of instruction. In the dry process, the nozzle handler has the responsibility of adding the correct amount of water at the gun position and clearly must have sufficient experience to do the job properly. Too much water will cause excess slump and the material will not adhere to the substrate, particularly on soffit and wall applications. If it is too dry, the rebound will be excessive.

Preparation of the surface to be sprayed is important, as a dirty or contaminated area will prevent adhesion. The surface can be cleaned by spraying with compressed air and water. If this is insufficient, it may be necessary to use sand-blasting. A soft substrate, e.g. sand, should always be sprayed first with a layer of grout before applying the concrete. This prevents segregation between the layer and the substrate.

Any water seepage or infiltration on surfaces should be sealed, using drainage channels if necessary. If water seepage is allowed to continue, there will be a loss of adhesion. On the other hand, a surface which is too dry will draw water out of the newly sprayed concrete. Such excessively dry surfaces should be well dampened beforehand, but care should be taken to ensure that no water is left on the surface when spraying begins.

Rebound is inherent in sprayed concrete work and is unavoidable. The rebound material is mainly the large aggregates and is most severe at the start of the work, particularly when spraying on to a hard surface. As the layer is built up, the rebound decreases because more material will adhere on impact. The amount of rebound will depend to some extent on the nature of the surface, but will depend mainly on the direction of the application. Overhead applications will obviously have more rebound than spraying down onto a surface (Fig. 147). The rebound quantities which may be anticipated vary from 0 to 10% for spraying downwards, up to 25% for vertical walls and increasing up to 40% for soffits. The quantities will differ, depending on the mix and other circumstances for specific jobs, but are always less if the wet process is used. Under no circumstances should any rebound material be reused, as quality control could be seriously affected.

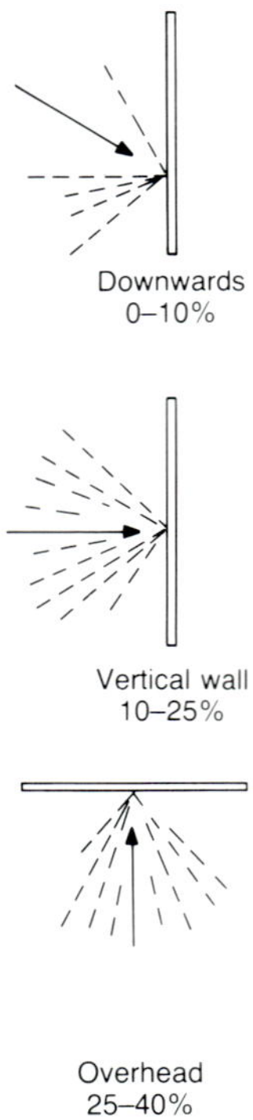

Fig. 147. Possible quantities of rebound

Mesh reinforcement size should be limited to 50 mm as

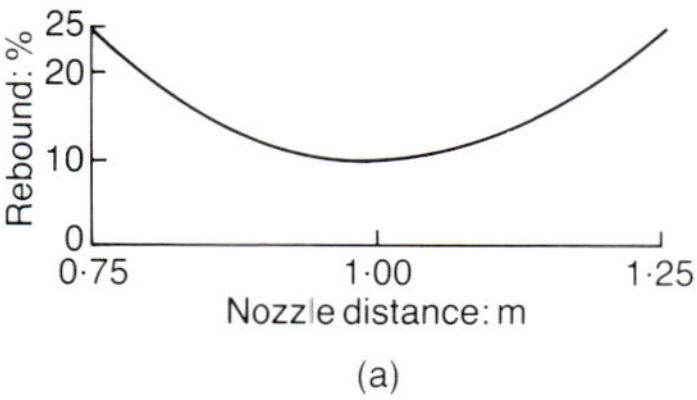

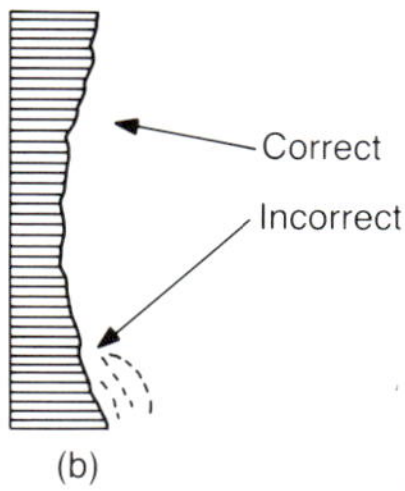

Fig. 148. Nozzle techniques: (a) distance and rebound; (b) spraying angle (courtesy Aliva Ltd)

smaller meshes will increase rebound. It is usual to spray an initial layer on to the substrate before placing the mesh. This makes it easier to fix the mesh, reduces rebound and ensures better adhesion. If there are two layers of mesh, the first one should be installed and completely covered with sprayed material before the second layer is fixed.

Preparing test cubes in the conventional manner is unsuit-

Fig. 149. Typical nozzle for wet process

able. Material making up a 100 mm thick layer is sprayed on to a rigid (non-vibrating) panel set up at the same angle as the work. After curing, 100 mm dia. cores are cut from a central position on the panel, capped and tested.

The most appropriate distance between the nozzle and the sprayed surface will depend mainly on the size and grading of aggregates and the surface finish required (Fig. 148). Normally the distance is about 1 m. If the nozzle is too near or too far away, the amount of rebound will increase. The nozzle handler is aiming for as little rebound as possible and also for a finished surface that meets the specification. The nozzle will be held at right angles to the sprayed surface for overhead work and slightly above the horizontal for vertical work. A typical nozzle for the wet process is shown in Fig. 149.

Safety

The HSE regulations[71] are naturally applicable to sprayed concrete as with any other construction work, but special precautions should be taken. It should be remembered that the technique involves working with compressed air and spraying concrete at high velocity on to a surface only a few feet away from the operator. That alone presents hazards, but there is also rebound to consider. Rebound material can reach speeds of up to 100 km/h and clearly the nozzle handler should wear protective clothing and/or equipment. Dust pollution is high, particularly with the dry process.

Protective clothing should include some form of helmet to completely cover the head, with a fresh air supply facility. The special safety considerations required for concrete pumping will also apply when the wet process is in use. Chapter 12 deals with safety and training in concrete pumping. Further reference can be made to the BCPA *Advisory safety code of practice for concrete pumping.*[6]

12

Safety and training in concrete pumping

In any branch of construction, safety and training are important and in this respect concrete pumping presents its own special hazards which arise principally from conveying concrete under pressure through a pipeline. Training operatives to work to well defined procedures will create benefits, not least being the reduction of accidents.

Many accidents have occurred in the past when cleaning out the pipeline. Recommendations are made in chapter 9 for cleaning out procedures.

12.1 The role of training in safety

When the subject of safety is talked or written about, training is usually included in the subject title, i.e. safety and training. Such a heading appears in countless reports and agendas for board meetings, albeit some perhaps well down in the agenda. It is suggested that even though the two go hand in hand, it should really be training and safety. This does not mean that training is more important than safety, but rather that a high standard of safety will normally follow on from a high standard of training. It is also suggested that firms who are placing this subject higher on the agenda will in due course reap benefits from it.

Much has been said and written about the intolerable safety record of the construction industry and how there are more fatalities and serious accidents to construction workers than in any other branch of industry. The very nature of the work makes it a hazardous occupation. There have been many laudable suggestions in the past as to the various ways in which accidents can be reduced — more safety officers, heavier penalties for offenders, stricter conditions of contract with emphasis on safety, more power for the engineer's staff on site, more willingness by site personnel to comply with safety instructions. The list is virtually endless.

It is beyond the scope of this chapter to elaborate on these various items, any one of which is important in its own right. It will be more appropriate to consider the training aspect and develop the theory that proper training will do more than anything else to promote safety. Even though the purpose of this chapter is to project the dual role of safety and training in concrete pumping, much of what will be said is still very applicable to the construction industry as a whole.

12.2 Why training?

There is no doubt that training should be part of a long-term plan in any business and not a spasmodic affair arising out of a situation which has arisen and which has demonstrated an urgent need for training. For example, a bad accident always makes people think of more training in safety matters, and the construction problem or mistake, due to the use of a faulty method, stimulates immediate thoughts of more procedural training.

The ICE published a simple but very effective education and training diagram which illustrates the advantage the trained person has in learning from a higher level than the untrained (Fig. 150). It is surely commercially advantageous for firms to do all they can to improve the quality of performance of the workforce at all levels. It has been proved that many people who have undergone training are more

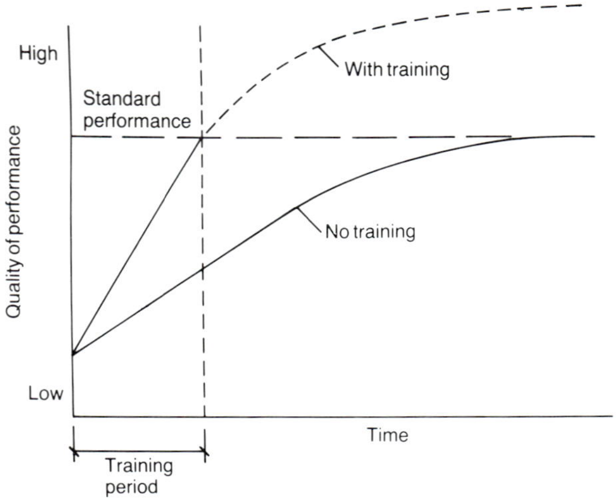

Fig. 150. Training leads to peaking faster (courtesy New Civil Engineer)

likely to reach a standard performance level much earlier in life than someone who has had no training. People continue to learn all their lives and the trained person will go on to higher performance levels much earlier in life than the untrained. Surely attaching much greater importance to training is sensible, not only in improving safety standards, but also commercially.

12.3. The main areas in training

As in any other part of the construction industry, adequate and proper training will provide considerable benefits. These will fall into four main areas: greater output and production; better use of plant and equipment; maintaining the quality of the end product; and a higher standard of safety.

Greater output and production

Using correct methods which have been carefully thought out to carry out the task logically, and using the resources available to the best advantage, will without doubt lead to increased output or production.

Better use of plant and equipment

Expertise in operating and maintaining plant and equipment will ensure proper use of these resources. The concrete pump operator may be in sole charge of £300 000 of plant, which is one good reason why the operator should be adequately trained.

Maintaining quality of the end product

In the case of concrete pumping, the end product is the in situ concrete and the expertise in conveying it to the pour position, then placing it properly, is of prime importance. Good quality material from the ready-mixed concrete truck-mixer or from the batching plant on site can be safeguarded.

Higher standard of safety

Training in the correct working procedures and use of plant will without doubt reduce accidents. This is a well known fact and is indisputable.

The commercial benefits to be gained from the four areas are obvious. The cynic may say that training costs money, but the answer to that is to do a rough calculation of the savings which can be made from increased output, reduced plant costs, reduced accident costs and the result will most certainly prove that training will pay for itself in the long term.

12.4. Training for concrete pumping

Many construction firms have realized the importance of producing written procedures (or instructions) relating to the various operations on site, ranging from menial tasks such as those carried out by the bricklayer's labourer to constructing an auger cast pile or erecting a tower crane. There is a correct way of doing a task: it is an advantage if the procedure is written down and used for training. Some firms, both large and small, have prepared training videos which assist in their training programmes. These procedures or films are prepared by people who have invariably learned the hard way — perhaps from bitter experience. They have been involved in solving problems and are well equipped to assist in the preparation of such procedures. Such carefully drawn up procedures are invaluable in the training of operatives in every branch of the construction industry. By following the rules the trainee is taught to carry out a task in a logical and safe manner. The trainee may be trained in house in both the classroom and training ground, but eventually the trainee will be carrying out these tasks on site and, it is to be hoped, following carefully these same procedures which the trainee has been taught.

It is virtually impossible to draw up a set of procedures for a task without becoming involved in the safety aspect of it. Writing down step by step procedures makes the writer aware of any safety hazards that may have been introduced. The opportunity is there to either work out another way which is safer, or if this is not practical to draw attention to the safety hazard. Committing a method or procedure of working to writing is extremely effective.

Concrete pumping has its own particular hazards and the BCPA has done much to promote safety by producing the *Manual and advisory safety code of practice for concrete pumping*,[6] which is the only publication of its kind in the world. It was drawn up with the assistance of the HSE, and the various aspects of safety in concrete pumping are dealt with in detail. In addition, most of the UK pump hire companies have their own procedure guides and instructions for their authorized (trained) operators.

There is evidence that more attention is being paid to

safety and training in the UK despite the fact that a higher percentage of total site placed concrete is placed by pump in other countries. Legislation will come into force in 1991 requiring mandatory formal pump operator training for all persons operating a concrete pump. BCPA members have been given the opportunity of advising and assisting the CITB in the preparation of training objectives.

As with any branch of the construction industry in the UK, all concrete pumping plant and equipment must be designed, manufactured and used in accordance with sections 5 and 6 of the Health and Safety at Work Act 1974, as amended by the EEC Product Liability Directive 85/374/EEC. The requirements of the Factories Act 1961 must also be observed.

12.5. Procedural methods and safety aspects of concrete pumping

Pumping concrete at high pressures obviously presents hazards and much of the procedural and safety training deals with this particular aspect of the work.

Pipelines and ancillary equipment

Pumping concrete through a pipeline presents a hazard in that the concrete is under pressure. It follows that the pipeline has to be strong enough to withstand the pressure. In section 2.2, the necessity of only using the correct size and wall thickness of pipeline is described in detail. Joint couplings, reducing pipes, pipe adaptors and hoses are all subject to the same pressure. Therefore, it is extremely important that this ancillary equipment is chosen with care. The information on correct sizes of pipeline, couplings, etc., will normally be obtained from the pump manufacturers, who recommend certain sizes for compatibility with the pump being used. Using a pump which is capable of producing a concrete pressure higher than that recommended for a particular size or wall thickness of pipeline is extremely dangerous. In such a situation there could be a serious accident.

Concrete pumping units

The manufacturers will normally supply, in the form of a manual, full details of the pump, and operating instructions should be carefully followed. A concrete pump is another item of construction plant and the relevant section of the Construction Regulations 1961,[73] i.e. general provisions on health and welfare, and working places, must be adhered to. Trailer or vehicle mounted pumps should be constructed in accordance with the Motor Vehicles (Construction and Use) Regulations 1973.[74] These are basic requirements.

Owning and operating a concrete pump presents certain additional hazards, many of which are related to the fact that the concrete is being moved under pressure. When using booms, either separately or with mobile pumps, stability is affected from time to time due to moments set up during pumping. The design of the pump can only go so far to counteract this: it is up to the user to ensure that the booms are stable. Correctly used radio remote control systems can be important aids to safety. The pump operator can be positioned at the best possible vantage point from where the operation can be seen. Operators claim that booms can be operated more sensitively using remote control. Such equipment can be used in certain circumstances without extra people for visual signalling, which can be confusing at times.

Hardstandings and use of stabilizers

All pumps should be positioned only on ground which is level and has been adequately prepared to take the weight of the pump and equipment. This area should include space for at least two truckmixers to back up to the rear of the pump, if this is possible. If pumps are equipped with outriggers, these are an integral part of the stability design and must be correctly used at all times. It is important that the outrigger foot rests on the ground which should be firm enough to withstand the pressure. It is usual to place steel plates or timbers underneath the outrigger foot (Fig. 151).

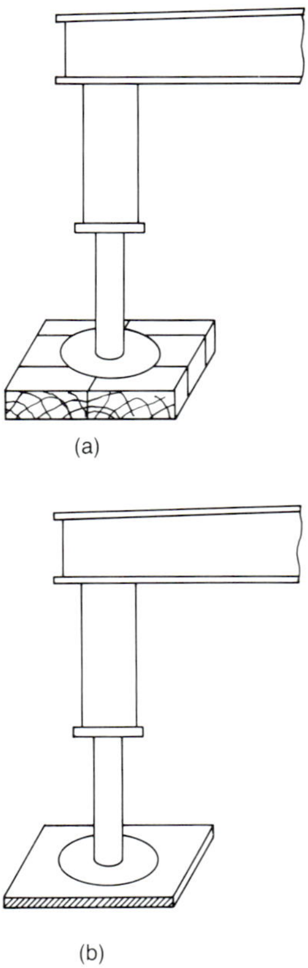

Fig. 151. Supports for outrigger foot: (a) three baulks laced together; (b) steel plate

With regard to machine stability when operating mobile pumps, Putzmeister has developed a new automatic stability control (ASC) which will automatically halt the boom movement if the stabilizers are not in the correct position, e.g. if the pump is not properly levelled. Such equipment is without doubt an important safety aid as serious accidents will occur if insufficient care is taken when setting up the pump.

Positioning a pump too near an open excavation is dangerous and, whether the excavation is supported or not, assurance should always be sought from the main contractor on this point. It is not possible to lay down fixed dimensions for the distance the pump should be from the side of the excavation. Much will depend on the depth of the excavation and the nature of the subsoil, but as a guide, if operating

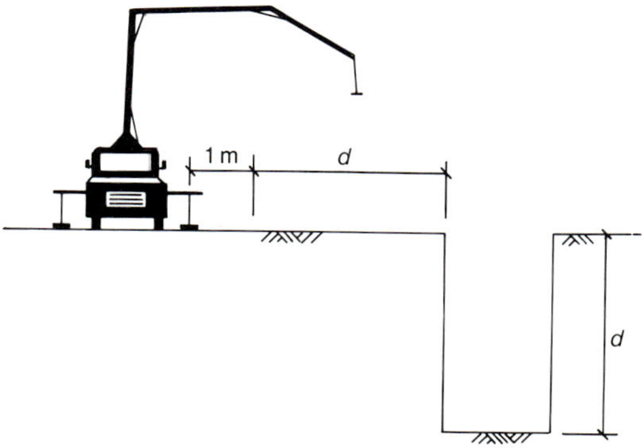

Fig. 152. Pump positioned close to an unsupported trench

near an unsupported trench, the near-side outriggers should not be nearer to the side of the trench than the depth of the trench plus 1 m (Fig. 152). Stability testing of concrete pump units should be carried out as specified in BS 1757.[74]

Dangerous parts of the pump unit

Any moving part of machinery is potentially dangerous and must be protected in accordance with BS 5304.[75] All concrete pumps will have a receiving hopper, within which there are rotating paddles or spades for remixing. In the past, many accidents have occurred mainly due to carelessness. If access is required to the hopper for any reason which entails lifting up the hopper grille, the shut-off valve should first be closed and locked, thus preventing the remixer shaft from rotating. No tool or part of the body should be inserted between the bars of the grille. The maximum spacing between bars should be about 60 mm.

Accidents have happened because a person has stood on the grille while remixing was taking place and the weight of the person was sufficient to deflect the grille. It is now recommended that the grille should be strong enough to withstand deflexion under the weight of a person and also should not be permanently deflected under the weight of the concrete being discharged into the hopper. The safety grille should be kept in position during the whole of the pumping operation.

Blow-backs may occur from the receiving hopper under certain circumstances and all personnel not actually concerned with backing the truckmixer up to the receiving hopper should keep clear of both the pump and the vehicle.

During this operation, the pump operator will normally act as banksman for the truckmixer and should not stand directly in front of the hopper but should stand well clear to one side.

Siting of a concrete pump

In areas used by the public, warning notices such as 'Keep clear — concrete pumping' should be placed in dangerous areas and, if necessary, members of the public excluded from the vicinity of the pump, boom and pipeline. Boom restrictions may require that the pump is sited so that adjacent buildings, scaffolding, etc., do not restrict the required boom movements. Sufficient clearance should be provided so that there is no risk of trapping personnel between the boom and the obstruction. Communications may present a problem if the pump is sited so that the pump operator is out of sight of the placing crew. In this event a proper system of signalling or other communication should be set up with a banksman appointed by the contractor so that there is no misunderstanding. As mentioned earlier, with long pipelines it may be necessary to use some form of radio communication.

12.6 Concrete pumping operations near high voltage cables

Special precautions should be taken when operating a mobile pump near overhead power lines. Fatal accidents can easily occur unless extreme care is taken. It should be remembered that a so-called flash-over can occur even without the boom touching the power line. It is recommended that the minimum safe working distance from the end of the maximum extended length of boom, measured with the boom pointing towards the power lines, is 8 m (Fig. 153).

In certain damp weather conditions the safe working distance may have effectively been reduced to a point where there is a danger of a flash-over. If such conditions arise, the electricity board or owners of the overhead lines should be contacted before the start of work. If for any reason work has to be carried out within the 8 m safety limit, the electricity supply must be cut off before starting work. A written permit of approval to work must also be obtained before the start of work, from the electricity board, the regional mechanical and electrical engineer in the case of British Rail, or a responsible person where other owners and operators of electrical power lines are concerned.

Even overhead domestic (220 V) cables can present a

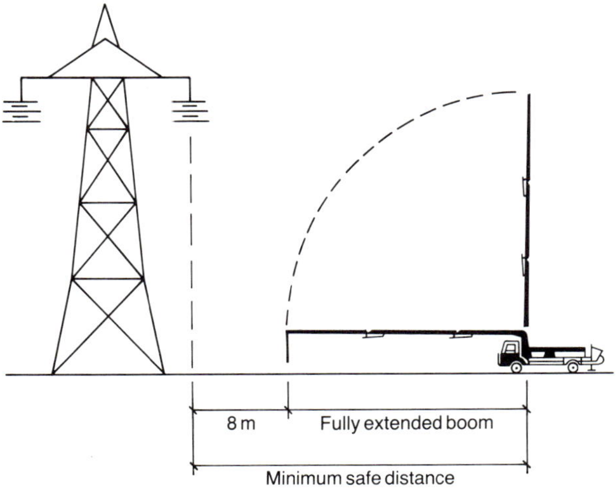

Fig. 153. Safe working distance from overhead HT cables

lethal hazard and fatal accidents have occurred under certain conditions, e.g. wet weather, when construction plant has broken a cable and the live end has come in contact with a worker. Workers must be aware of the danger from overhead electrical cables. This awareness should be part of the training of all pump operators. If any part of the pump is in contact with an HT cable, the operator should remain in the same position on the pump, tell others to keep clear and to inform the electricity board. If the operator cannot disengage the pump from the power line, the operator must stay put. If it is absolutely necessary to jump clear, the operator must not touch the ground and the pump at the same time.

Section 44 of the Construction (General Provisions) Regulations 1961[72] should be read in conjunction with this section.

12.7. Safety check-list for concrete pumping

1. Remember that concrete is being conveyed under pressure.

2. Assume that any part of the pipeline is under pressure until the pump operator has said that the pipeline is not under pressure.

3. No pipe coupling should be released or work carried out on the pipeline until the pressure has been released by reverse pumping.

4. Remember that a blockage can result in very high pressures in the pipeline.

5. Cleaning the pipeline using compressed air should only be carried out under the supervision of an experienced person. The person must have an experienced assistant.

6. The pipeline should only be cleaned using compressed air in circumstances where it is not practical to use high pressure water.

7. All instructions and precautions laid down for cleaning using compressed air should be carried out exactly, as the operation can be extremely dangerous.

8. A catch basket fixed to the end of the pipeline and wash-out adaptor must be used for all cleaning operations.

9. Pumps should be carefully sited in operations near high voltage power lines.

10. On restricted sites, mobile pumps with booms must be positioned so that when the boom is slewed people are unlikely to be trapped by the boom.

11. Pipe thicknesses should be chosen with care to match the pump being used.

12. Worn pipes or joints could burst under concrete pressure and cause serious injuries. Excessively worn equipment should not be used.

13. Quick-acting toggle type couplings must have a safety pin attached, to prevent accidental release or loss of the pin.

14. Placing hoses should be handled with care and kinking should be avoided, as this can result in whiplash.

15. Placing crews should avoid working under a placing boom.

16. Special care should be taken when working in public places and 'Keep clear' notices should be posted or the public excluded from the working area.

17. All equipment should be stowed securely and the boom folded into position when travelling with a mobile pump.

18. Always check the siting of the pump so that it is not too near an excavation, drains or manholes.

19. Never stand between the pump and the truckmixer when the latter is reversing up to the pump hopper.

20. When entering a site in the dark (pumps often arrive early in the morning), proceed with care. Preferably inspect the route on foot to check that it is suitable and safe.

21. Do not site the mobile pump on an incline, particular-

ly sloping downwards in the direction of the boom, as this could lead to instability during pumping.

22. The mobile pump boom should never be used for lifting personnel or equipment or for dragging hoses or pipelines along the ground.

12.8. Summary

All construction sites are potentially dangerous places in which to work, some more so than others depending on the nature of the work. Concrete pumping has its own special hazards, which arise principally from working with materials which are being moved at high pressure. The use of high pressure can sometimes create circumstances which cause stoppages of work, e.g. blockages. The blockages entail remedial measures to rectify. Carrying out these remedial measures can be dangerous if certain procedures are not followed carefully. It follows that these procedures must be laid down in detail and the workforce trained to carry them out. Formal training of all pump operatives including placing crews is therefore necessary if this work is to be carried out properly. In doing so, companies will be creating a base for greater safety awareness and for fewer accidents.

Training at all levels in many construction companies is on the increase, but as yet much of it takes place as something extra to the essential planning aspects of company growth. In reality it should be considered as central to business and the costs should be allowed for as such. Investment in training can be a sound investment which will produce the skills necessary to compete in a highly competitive industry. Safety and training go together, but if training is started with, safety will assuredly follow.

Appendix 1. Main manufacturers of concrete pumps and equipment

It is not practical to provide a comprehensive list of all the manufacturers of concrete pumps throughout the world and this appendix will cover only some of the main manufacturers. Many of the earlier manufacturers of concrete pumps no longer exist, but names such as Torkret (Germany), Almacoa (France), Winget (UK), Ransomes and Rapier (UK), Challenge Cook Bros (USA), and Sidewinder (Australia) will always be associated with concrete pumping. Other later companies such as Mark-Thomsen (USA) and Wibau (Germany) have been taken over by other manufacturers. The world leaders are the German firms Schwing and Putzmeister. Consequently much of this appendix will concern these two organizations.

The names Schwing and Putzmeister are synonymous with concrete pumps. The rivalry between these two companies has without doubt created considerable benefits for the concrete pumping industry throughout the world. Both firms have contributed greatly to the development of pumps and equipment and the competitive spirit between them can rightly be said to have done much to produce virtually trouble-free concrete pumps. To

*Table 30. Schwing mobile unit construction system**

Parameter	Concrete pump model				
	650	601HD	801	900HDR	1200HDR
Maximum theoretical concrete output: m^3/h	62	66	82	90	130
Maximum concrete pressure: bar	55	70	57	108	96
Placing boom model	KVM21	KVM23	KVM23-4H	KVM26	KVM28
Vertical reach: m	20·8	22·7	22·7	25·5	27·7
Horizontal reach: m	17·55	19·45	19·05	21·9	24·0
Placing boom model	KVM31	KVM32	KVM36	KVM42	KVM52
Vertical reach: m	30·75	31·62	35·8	41·65	51·2
Horizontal reach: m	27·0	28·0	32·1	38·05	48·0

* Concrete delivery pipeline diameter is 125 mm throughout.

Fig. 154. Mobile pump with 42 m boom (courtesy F. W. Schwing/Burlington Engineers)

tour the factories of these two market leaders in the production of concrete pumps cannot fail to impress the visitor, not only in production methods being used, but in the vast resources employed in design, development and in the rigorous testing which is carried out as a matter of routine.

Schwing concrete pumps are distributed in over 120 countries. Friedrich Wilhelm Schwing began making construction equipment in 1934 and after the Second World War turned to developing and manufacturing concrete pumps. Schwing's first development was the pneumatic placer and a twin cylinder hydraulic concrete pump was marketed in 1957. Since then the company has made rapid progress: Friedrich Schwing is an ardent supporter of the saying 'standing still is back-sliding'. His sons are now joint managing directors of the company. Schwing's main factory is at Herne, in

*Table 31. Schwing range of portable concrete pumps**

Portable concrete pump model	Maximum drive power: kW	Maximum theoretical concrete output: m^3/h	Maximum concrete pressure: bar
BP750RE	22	35	65
BP750R	38	55	60
BP1000R	51	53	60
BP250HD	49	32	87
BP550HD	111	66	70
BP550HD-18R	111	70	70
BP3000HD	132	87	110
BP3000HD-18R	160	73	135
BP5000HD	2 × 111	135	135

* Output and pressure figures vary on piston side and rod side drive: maximum figure is shown here.

Fig. 155. Roll 'n' fold boom system (courtesy F. W. Schwing/Burlington Engineers)

West Germany, and the company has manufacturing branches in other parts of the world including the USA.

The Schwing mobile pump range consists of five models systematically designed on a modular construction principle. The modular construction facilitates combination with appropriate booms and pipelines, thus providing the most suitable and economical units (Table 30). There are ten different boom combinations, two suitable for each size of pump and are constructed in three or four sections (Fig. 154). Schwing's roll'n'fold design for the boom folded position is compact and neat, even on the very large

Fig. 156. BP250 multi-purpose concrete pump (courtesy F. W. Schwing/Burlington Engineers)

Fig. 157. BP5000HD stationary/trailer pump (courtesy F. W. Schwing/Burlington Engineers)

units (Fig. 155). The original Schwing folding arrangement is now used extensively by other manufacturers.

There are nine models in the portable (trailer/stationary) range of Schwing's concrete pumps (Table 31). Two of the models are shown in Figs 156 and 157. Schwing also manufactures truckmixer pumps (two models), and a wide range of tunnel pumps equipped with belt conveyors and/or mixing equipment (Fig. 158). The

Table 32. Basic Putzmeister mobile pumps incorporating the elephant trunk transfer tube system

Parameter	Truck mounted concrete pump						
	BF 1405	BRF 1406	BRF 1408	BRF 1411	BRF 2112	BRF 2100H	BRF 2115
Effective output: m^3/h	50	60	80	110	120	Up to 150	150
Concrete pressure: bar	85	70	54	54	54	Up to 140	85
Cylinder diameter: mm	180	200	230	230	230	180/200/230	230
Stroke: mm	1400	1400	1400	1400	1400	2100	2100

Fig. 158 (facing page). Truck-mixer concrete pump (courtesy F. W. Schwing/Burlington Engineers)

Fig. 159. BRF2112 mobile pump with 62 m boom (courtesy Putzmeister Ltd)

Fig. 160. Multi-purpose BSA 1004 trailer pump (courtesy Putzmeister Ltd)

company produces four trailer types with booms specially designed for tunnel work.

Putzmeister GmbH is based near Stuttgart in West Germany but also has factories in other parts of the world. The firm started producing mobile pumps in 1968 using a flapper valve and since then has had a continuing programme of development: the latest mobile pumps use the Elephant trunk system. These pumps range from the BF1405 rated at 50 m^3/h to the huge BRF2112 rated at 150 m^3/h (Table 32). The BRF2112 is fitted with a five-section

Table 33. Putzmeister basic range of trailer pumps incorporating the S valve system (courtesy Putzmeister Ltd)

Pump model	Maximum theoretical output: m^3/h	Concrete pressure: bar*	Stroke: mm	Cylinder dia.: mm	Concrete hopper: l	Diesel engine: kW	Electric motor: kW	Frame/type of hood
BSA 1002	20	75	1000	120	330	28	30	C
BSA 1003	35	46	1000	180	400	37	30	C
BSA 1004	40	70	1000	180	600	56	45	F
BSA 1406	60	70	1400	200	600	75	75	F
BSA 1408	80	70	1400	200	600	118	90	F
BSA 2112H	120	115	2100	200	600	177	160	H

* For some models, pressure can be increased by 30% by feeding oil to the cylinder heads.

Table 34. Putzmeister basic range of trailer pumps incorporating elephant trunk valve system (courtesy Putzmeister Ltd)

Parameter	Pump model			
	BA 1405	BRA 1406	BRA 1400 H	BRA 2100 H
Effective output: m^3/h	50	60	80	90
Concrete pressure: bar	85	70	100	115
Cylinder dia.: mm	180	200	200	200
Capacity: kW	75	118	118	175
Weight: kg	approx 3100	4500	5300	5600

boom, giving a vertical placing off-the-boom height of 62 m (Fig. 159).

The company also manufactures trailer pumps using the S valve (Table 33) and the elephant trunk arrangement (Table 34). The smaller trailer pumps (Fig. 160) in these ranges are multi-purpose and are widely used not only for normal concrete, but for floor screeding, sprayed concrete and tunnel applications. The larger pumps have produced world records for height and distance and the BSA2112H holds the current world record for pumping to a height of 432 m (see chapter 2). Putzmeister's largest trailer pump

Fig. 161. PTS 14000 (S valve) trailer pump (courtesy Putzmeister Ltd)

Fig. 162. City concrete pump (courtesy Putzmeister Ltd)

Fig. 163. Pumi truckmixer concrete pump (courtesy Putzmeister Ltd)

Fig. 164. Mobile pump with 32 m boom (courtesy Elba-Werk GmbH)

is the PTS14000 with an output of up to 150 m^3/h and a concrete pressure of up to 138 bar (Fig. 161).

The firm has introduced a new unit recently which is called the City Pump. This pump is designed for use where access and space are problems. It is compact, manoeuvrable, and is available as a trailer pump on a 7·5 t chassis or as a mobile pump with a 16 m boom mounted on a 10 t chassis. The effective output of the pump is 40 m^3/h (Fig. 162). Putzmeister's full manufacturing range includes truckmixer pumps (Fig. 163) and a range of pneumatic pumps including the Mixokret for floor screed conveyance (see chapter 10).

Other West German manufacturers include the long established Elba-Scheele GmbH, who also produce a full range of mobile, stationary and tunnel pumps together with concrete mixing plants and truckmixers. The headquarters are at Ettlingen in West Germany and there are branches in most parts of the world. The company claims that its pumps will handle extremes in concrete mixes (harsh aggregates) with ease and that its pumps are simply constructed and extremely robust (Fig. 164).

Wibau is another West German manufacturer, now part of Putzmeister, which was originally well known as a producer of peristaltic pumps. Although this type of pump has not been so widely used in recent years, many are still in operation, which is a tribute to the design. Fig. 165 shows a typical Wibau peristaltic pump.

In the UK there are now several manufacturers of concrete pumps. Mokan Ltd of Southwell, Nottinghamshire, produce and

Fig. 165. Peristaltic BRP80 mobile pump with 32 m boom (courtesy Wibau GmbH)

distribute small–medium sized trailer pumps, which are built around its Ackermann A valve. This company was first in the UK in the design and manufacture of low output trailer mounted pumps in 1977 and went on to develop further skid mounted pumps, grout and plaster pumps for mining and underground work generally. The company manufactures two sizes of trailer mounted pumps, the P25 and P40, with outputs of 25 and 40 m^3/h, respectively (Figs 166 and 167). A large fleet of these pumps is owned and hired by an associate concrete pump hire company, Royston Conpump Ltd.

Fig. 166. Mokan P25 Ackpump trailer concrete pump (courtesy Royston Conpump Ltd)

Fig. 167. Mokan P40 Ackpump trailer concrete pump (courtesy Royston Conpump Ltd)

Mokan also mounts these pumps on a chassis with a 17·5 m long, three-section placing boom are fitted. The skid mounted tunnel and mining pumps are contained within a volume of 0·9 m × 0·94 m × 2·5 m. For placing concrete in areas which are inaccessible to the larger mobile units, Mokan recently mounted its P25 pump on a small 7·5 t chassis with a 10 m boom (Fig. 168). Mokan claims that this pump is tailor made for refurbishment work. The company has plans to mount the P40 in a similar fashion.

Hydroseal Concrete Pumps Ltd (part of the AFD Group of

Fig. 168. P25 mobile pump with 10 m boom (courtesy Mokan Ltd)

Table 35. Hydroseal data

Pump	Output
Mini (static/trailer)	35 m^3/h up to 70 bar 55 m^3/h up to 45 bar
Midi (static/trailer or mobile)	56 m^3/h up to 75 bar
Maxi (static/trailer or mobile)	70 m^3/h up to 70 bar 110 m^3/h up to 45 bar

companies), who are based in London, are the only British designers and manufacturers of both trailer and mobile pumps. Development has proceeded in conjunction with an associate pump hire company, AFD Concrete Pumps Ltd. Hydroseal has been able to make use of first-hand information from this company which has had many years of experience in a wide variety of situations. The firm is the second largest concrete pump hire company in the UK.

Hydroseal has three standard models, the Mini, Midi and Maxi (Table 35). The Mini is a static or trailer unit, but the other two can be supplied as mobile units (Figs 169 and 170). Boom lengths vary from 16 m to 36 m with 100 or 125 mm dia. pipelines. The mobile units can be mounted on most two, three or four-axle chassis/cab units.

Sidewinder Concrete Pumps Ltd, based in Scotland, is involved

Fig. 169. Mobile concrete pump (courtesy Hydroseal Concrete Pumps Ltd)

in the manufacture, sale and hire of a variety of trailer and truck mounted concrete pumps. The company also has a wide range of grout, mortar, screed and plaster pumps. Sidewinder is one of the oldest concrete pumping companies in the UK; the original Sidewinder pump originated many years ago in Australia. The Sidewinder SW range of trailer pumps has theoretical outputs up to 50 m^3/h (Table 36).

Fig. 170. Maxi trailer concrete pump (courtesy Hydroseal Concrete Pumps Ltd)

These pumps are recognized for their reliability, and ease of operation and maintenance. The concrete valve is a simple swinging tube arrangement which gives a direct flow of concrete from hopper to cylinder and from cylinder into the pipeline. The company provides a fully enclosed sound attenuated engine cover which controls noi

Table 36. Sidewinder

Pump	Output: m^3/h	Line pressure: bar	Engine
SW 30	30	60	80 HP, 4 cylinder water-cooled diesel
SW 40	40	45	80 HP, 4 cylinder water-cooled diesel
SW 50	50	60	120 HP, 4 cylinder turbocharged diesel

* Optional electric prime movers are available.

Fig. 171. SW30 hushpump trailer concrete pump (courtesy Sidewinder Concrete Pumps Ltd)

construction equipment. The SW30 model is particularly popular with continuous flight auger piling companies (CFA piling) in the UK (Figs 171 and 172).

Transcrete Sidewinder trailer concrete pumps are manufactured in Australia and under licence in the USA and Japan. They are distributed in the UK by Transcrete (UK) Ltd, based at Sherridge, Worcestershire. These trailer pumps have been in continuous production for many years, yet still maintain the original concept. The company produces four standard models with theoretical outputs of 30, 35, 45 and 55 m^3/h. The Sidewinder 45 is the most popular model. Although these pumps are normally classed as trailer types (Fig. 173) they can be truck or skid mounted.

Fig. 172. SW40 trailer concrete pump (courtesy Sidewinder Concrete Pumps Ltd)

Fig. 173. Typical Transcrete Sidewinder trailer concrete pump (courtesy Transcrete (UK) Ltd)

Fig. 174. CFA piling application for Sidewinder concrete pump (courtesy Transcrete (UK) Ltd)

The Transcrete Sidewinder models are another example of simple, reliable, small–medium size concrete pumps and are used for many applications, not only conventional construction work but for tunnelling, shotcreting, grouting and continuous flight auger piling (Fig. 174). The main features of the Transcrete Sidewinder are the transfer tube valve system, a smooth soft push hydraulic system and piston/cylinder wear parts which the company claims will last for many years with sensible and careful handling.

The Italian company Italmacchine SpA, (UK agents: Utranazz (Bourne End) Ltd), is part of the CIFA SpA group and is based near Perugia. Italmacchine is a manufacturer of medium output pumps and associated equipment. The pumps are rated from 15 m^3/h to 45 m^3/h and will handle moderate quantities of concrete without having to use larger, more costly pumps. Three models with outputs from 20 to 60 m^3/h are available in the trailer range, with or without booms and use a transfer tube system which Italmacchine calls a rocking tube valve. The two or three section booms have a vertical reach of 10–18 m and a standard rotation of 350°.

These relatively small pumps with optional boom attachments will undoubtedly interest the contractor who specializes in smaller pours and does not wish to buy more expensive equipment than is absolutely necessary. The firm also produces a truck mounted unit equipped with six different types of booms ranging from 17 m to 25 m vertical reach (Fig. 175). Another innovation from Italmacchine is the Spider separate placing boom mounted on a one axle trailer with 15 m vertical and 9 m horizontal reaches. Any of the pumps can be connected to this separate placing boom (Fig. 176).

Fig. 175. Hydromix BR23/25 mobile concrete pump (courtesy Utranazz (Bourne End) Ltd)

Fig. 176. Hydromix BR445 trailer concrete pump fitted with separate placing boom (courtesy Ultranazz (Bourne End) Ltd)

Fig. 177. CIFA 8 series portable concrete pump (courtesy CIFA SpA)

CIFA SpA, part of the CIFA group, is based at Novate Milanese, near Milan, and is a leading European manufacturer of all types of equipment for concrete production which includes concrete mixing plants, truckmixers, concrete pumps, steel formwork and associated equipment. The firm is one of the pioneers of climbing steel formwork, which was established in 1928. During the past decade, CIFA's technology has been concentrated on concrete pumping and the module developed has produced a wide range of modern concrete pumps.

The concrete pumps have maximum theoretical outputs ranging from 20 m^3/h to 130 m^3/h. The stationary models can be mounted on trucks and fitted with placing booms which have vertical reaches of 17–40 m. CIFA uses a simple S type valve with a straight through flow and claims good sealing and wear qualities (Figs 177 and 178).

CIFA has two versions of the Pompuno, which is a truckmixer concrete pump with maximum theoretical outputs of up to 65 m^3/h and vertical reaches of 17–23·7 m. Mixer sizes are 6, 7, 8 and 9 m^3. An S type valve is again included in the design and the pumping module has similar characteristics to the other CIFA concrete pumps. The unit can be fitted to any make of chassis (Fig. 179).

The company also produces separate placing booms in five sizes with horizontal reaches of 17–33·6 m and the booms can be interchanged with the boom types used in the truck mounted series. Other

Fig. 178. CIFA 8 series Aconcagua mobile pump (30/28)(courtesy CIFA SpA)

Fig. 179. Pompuno truckmixer concrete pump with shotcrete boom fitted (courtesy CIFA SpA)

equipment manufactured includes tunnel pumps and shotcrete pumps for the wet process. The latter can be portable units or trunk mounted with three-section booms. They are particularly suitable for spraying concrete in tunnel construction.

In other parts of the world, the markets are generally dominated by Schwing and Putzmeister, but one exception is Japan. Kyokuto is the largest Japanese manufacturer: about 50% of production is small mobile pumps of 6–8 t gross weight and boom lengths of 10–15 m. Originally the Japanese made squeeze type pumps but have now turned to piston pumps with transfer tube valve systems. Niigata Engineering Co. Ltd has a range of medium sized mobile pumps with outputs up to 95 m^3/h and boom lengths of just over 20 m. This company also produces trailer pumps.

There are several manufacturers in the USA, but many now operate in conjunction with the leading West German manufacturers. Some of the long-standing manufacturers still produce a wide range of concrete pumps. Such firms include the Morgen Manufacturing Co., Yankton SD, Mayco Pump Co. in Los Angeles, and the Allentown Pneumatic Gun Inc. who still make trailer and mobile pumps as well as the shotcrete equipment for which the company is best known. The manufacturing companies of Thomsen and Pecco are now part of Putzmeister, who has a very large share of the market.

Australian manufacturers are located mainly in the Sydney area and produce medium sized pumps such as the Sidewinder range; 90% of these are mounted on a mobile chassis and are called mobile pumps but only 40% have booms. Thus there is a relatively low number of trailer pumps in Australia. Manufacturers include Seaboard Western Corp., Pty., Jacon Engineering, Flowcrete and Powercrete. Rawlston Schwing import and assemble Schwing pumps.

References

1. ANSON M., ASTON D.A. and COOKE T.H. *The pumping of concrete — a comparison between the UK and West Germany.* University of Lancaster/Polytechnic of Wales, 1986.
2. FITZGIBBON M.E. Large pours for reinforced concrete structures. CPS 28. *Concrete*, March 1976.
3. HMSO. *Construction (Lifting Operations) Regulations 1961*, HMSO, London, 1961.
4. BRITISH STANDARDS INSTITUTION. *Code of practice: safe use of cranes.* BSI, London, 1972, CP 3010.
5. ILLINGWORTH J.R. *Site handling equipment.* Thomas Telford, London, 1982, 60, 61.
6. BRITISH CONCRETE PUMPING ASSOCIATION. *Manual and advisory safety code of practice for concrete pumping.* BCPA, 1988, rev. edn.
7. EDE A.N. The resistance of concrete pumped through pipelines. *Mag. Concr. Res.* 1967, **27**, Nov, 129–140.
8. WEBER R. *The transport of concrete by pipeline.* Cement and Concrete Association, London, 1968, library translation 129.
9. LOADWICK F. Some factors affecting the flow of concrete through pipeline. *Proc. 1st Int. Conf. Hydraulic transport of solids in pipes.* British Hydromechanics Research Association, Bedford, 1970, D1–32.
10. BROWNE R.D. and BAMFORTH P. Tests to establish concrete pumpability. *J. Am. Concr. Inst.*, 1977, **74-19**, May, 197–202.
11. KEMPSTER E. Measuring void content. *Contract J.*, 1969, **229**, 27 Mar., 409–410.
12. CONCRETE SOCIETY. *Pumping concrete*, CS, London, 1984, digest 1, 3.
13. KEMPSTER E. *Pumpable concrete*, Building Research Station, Watford, 1969, current paper 29/69.
14. VON ECKARDSTEIN E. *Pumping concrete and concrete pumps.* F.W. Schwing, Herne, West Germany, 1983.
15. AMERICAN CONCRETE INSTITUTE. *Placing concrete by pumping methods.* ACI, Detroit, 1982, report 304-2R-71, rev. edn.
16. AMERICAN CONCRETE INSTITUTE. *Proportioning concrete mixes.* ACI, Detroit, 1974, report SP-46.

17. BRITISH STANDARDS INSTITUTION. *Code of practice: concrete admixtures*. BSI, London, BS 5075 Part 1: 1982; Part 2: 1982; Part 3: 1985.
18. CEMENT AND CONCRETE ASSOCIATION/CEMENT ADMIXTURES ASSOCIATION. *Superplasticising admixtures in concrete*. C&CA/CAA, Wexham Springs, 1976, Joint Working Party Report 45.030.
19. BRITISH STANDARDS INSTITUTION. *Structural use of concrete*. BSI, London, 1985, BS 8110.
20. DANISH MINISTRY OF TRANSPORT, ROAD DIRECTORATE. *The concrete of the Faro bridges*. MoT, Denmark, 1985.
21. GAYNOR R.D. Control of air content in concrete. *American Concrete Institute Convention, Mar. 1980*. ACI, Detroit, 1980.
22. FLETCHER K.E. *Admixtures for concrete*, Building Research Establishment, Watford, 1974, current paper 3/74.
23. CEMENT ADMIXTURES ASSOCIATION. *Admixtures data sheet*. CAA, Southampton, 1985, 7th edn.
24. HEWLETT P.C. *Cement admixtures: uses and applications*. Longman Scientific and Technical, Harlow, 1988, 2nd edn.
25. BRITISH STANDARDS INSTITUTION. *Code of practice: pulverized fuel ash for use in concrete*. BSI, London, BS 3892, Part 1: 1982; Part 2: 1984.
26. BRITISH STANDARDS INSTITUTION. *Code of practice: ground granulated blastfurnace slag*. BSI, London, 1986, BS 6699.
27. CEMENT AND CONCRETE ASSOCIATION. *Report on use of PFA and GGBFS*. C&CA, Wexham Springs, 1986.
28. DEPARTMENT OF TRANSPORT. *Specification for roads and bridges*. HMSO, London, 1986.
29. CARBRERA J.G. and WOOLEY G.R. A study of 25 year old pulverized fuel ash concrete used in foundation structures, *Proc. Instn. Civ. Engrs*, Part 2, 1985, **79**, 149–165.
30. WOOLEY G.R. and GAINS J.G. Precast prestressed piling at Drax Power Station and the use of pulverized fuel ash replacement in concrete. *Proc. Instn Civ. Engrs*, Part 1, 1984, **76**, May, 473–495; discussion: 1985, **78**, Aug., 1001–1008.
31. BRITISH STANDARDS INSTITUTION. *Code of practice: Portland blastfurnace cement*. BSI, London, 1973, BS 146.
32. BRITISH STANDARDS INSTITUTION. *Code of practice: low heat Portland blastfurnace cement*. BSI, London, 1974, BS 4246.
33. BLUNDELL R. and BROWNE R.D. Large pours for RC structures. *Proc. One-day Symp. University of Birmingham, 25 Sept. 1973*, Concrete Society, West Midlands Region, 1973.
34. BUILDING RESEARCH ESTABLISHMENT. *Concrete in sulphate bearing soils and ground waters*. HMSO, London, 1981, BRE digest 250.
35. CONCRETE SOCIETY. *Alkali silica reaction. Minimizing the risk of damage to concrete*. CS, London, 1977, technical report 30.
36. CONCRETE SOCIETY. *Microsilica concrete. Part 1: The material*. CS, London, 1986, current practice sheet 104.
37. CONCRETE SOCIETY. *Microsilica concrete, Part 2: In use*. CS,

London, 1986, current practice sheet 110.
38. Henderson J. *Influence of silica fume on the durability of concrete.* Elkem Chemicals, High Wycombe, 1986.
39. British Board of Agrement. *Agrément Certificate 85/1568.* BBA, 25 September 1986.
40. Henderson J. *The cementing efficiency of microsilica in British Portland cement concretes using gravel aggregates.* Elkem Chemicals, High Wycombe, 1986.
41. Henderson J. *The cementing efficiency of microsilica in British Portland cement concrete using limestone aggregates.* Elkem Chemicals, High Wycombe, 1986.
42. Institution of Structural Engineers. *Guide to the structural use of lightweight aggregate concrete.* ISE, London, 1988.
43. Concrete Society. *Comparing the costs of lightweight concrete and dense concrete.* CS, London, 1983.
44. Pozzolanic Lytag. *Technical bulletin 2.* PL, Hemel Hemstead, 1985.
45. Quality Scheme for Ready Mixed Concrete. *Manual of quality systems for concrete.* QSRMC, Walton on Thames, 1984.
46. British Standards Institution. *Quality systems.* BSI, London, BS 5750, Parts 0–6: 1981–87.
47. Dewar J. and Anderson R. *Manual of ready mixed concrete.* Blackie, Glasgow and London, 1988.
48. Anson M. and Cooke T.H. *The operation and capacity of the ready mixed concrete industry — a regional study.* University of Lancaster, 1988.
49. British Standards Institution. *Code of practice for maritime structures,* BSI, London, 1984, BS 6349.
50. Concrete Society. *Underwater concreting.* CS, London, 1971, technical report 3.
51. Session 5B. Geotextiles Special Applications. *2nd. Int. Conf. Geotextiles. Las Vegas, 1982.*
52. Lupton T.J. The advancing techniques in flexible forms. *2nd Int. Conf. Geotextiles, Las Vegas, 1982.*
53. Thomas Telford Ltd. CFA problems needs study. Report on Ground Engineering Group Meeting. *New Civ. Engr,* 19 Nov. 1987, 59.
54. Thomas Telford Ltd. Piling without the guesswork. *New Civ. Engr,* 24 Mar. 1988, 22.
55. Institution of Civil Engineers. *Specification for piling,* ICE, London, 1988.
56. Illingworth J.R. *Movement and distribution of concrete.* McGraw-Hill, Maidenhead, 1972, 18.
57. Illingworth J.R. Planning for the pump. *Concrete,* 1982, Nov., 20, 21.
58. Fitzgibbon M.E. Large pours for reinforced concrete structures. CPS 28. *Concrete.* 1976, Mar., 81.
59. Fitzgibbon M.E. Large pours heat generation control. CPS 35.

Concrete, 1976, Dec., 103–105.
60. FITZGIBBON M.E. Large pours continuous casting. CPS 36. *Concrete*, 1977, Feb., 107, 108.
61. ILLINGWORTH J.R. Getting the sums right. *1st Int. Conf. Ready mixed concrete in the UK, Dundee University, Sept. 1975.*
62. DRESSEL K.-M. and OSWALD F. *Improvement of concreting work.* Institute for the Science of Labour and Construction Management, Bonn, 1984.
63. RYAN T.F. *Gunite: a handbook for engineers.* Cement and Concrete Association, Wexham Springs, 1973.
64. AMERICAN CONCRETE INSTITUTE. *Shotcreting.* ACI, Detroit, 1966, SP-14.
65. HENDERSON J. *The repair and protection of concrete structures.* Concrete Constructions Exhibitions, Lutterworth, 1988, 44, 45.
66. ILLINGWORTH J.R. *Temporary works—their role in construction.* Thomas Telford, London, 1987.
67. SPRAYED CONCRETE ASSOCIATION. *Fibre reinforcement techniques data sheet 1.* SCA, London, 1987.
68. AMERICAN CONCRETE INSTITUTE COMMITTEE 506. *Guide to shotcrete.* ACI, Detroit, 1985, 506R-85.
69. AMERICAN CONCRETE INSTITUTE. *Application and use of shotcrete.* ACI, Detroit, 1981, ACI compilation 6.
70. SPRAYED CONCRETE ASSOCIATION. *Guidelines and specification for sprayed concrete.* SCA, London, 1986.
71. HMSO. *Health and Safety at Work Act 1974.* HMSO, London, 1974.
72. HMSO. *Construction (General Provisions) Regulations 1961.* HMSO, London, 1961.
73. HMSO. *Vehicle (Construction and Use) Regulations 1973.* HMSO, London, 1973.
74. BRITISH STANDARDS INSTITUTION. *Stability testing of concrete pumping units.* HMSO, London, 1986, BSI 1757.
75. BRITISH STANDARDS INSTITUTION. *Protection of moving parts.* BSI, London, 1975, BS 5304.

Glossary

accelerator. Admixture which, when added to concrete, increases the rate of hydration, shortens the setting time, or increases the rate of hardening or strength development.

admixture. Substance, other than aggregate, cement and water, which is added to the concrete mix to alter its properties.

aggregate. Broken or crushed stone, gravel, sand or similar inert material which forms a substantial part of concrete.

air entrainment. Inclusion of air in the form of minute bubbles during the mixing of concrete.

air meter. Device for measuring the quantities of air entrained in concrete.

air pockets. Pockets of air trapped in the pipeline or in the pump valve housing.

alkali-aggregate reaction. The chemical reaction in concrete between alkalis (sodium and potassium) released from Portland cement or other sources.

anchoring (of pipeline). Fixing the pipeline to the ground or to a structure.

angular movement. Movement which can take place in a pipeline because of the coupling design.

automatic mast control. Control which automatically monitors stability of mobile pumps.

A valve. Type of valve manufactured by Mokan Ltd.

bends (pipeline). Short pieces of pipeline in various radii for changing the direction of the pipeline.

blank flange. Metal flange attached to the end of a pipeline and has a drain cock attached.

blast-furnace slag (granulated). Glassy granular material formed when molten blast-furnace slag is immersed in water.

bleeding. Water leakage from the mix causing disintegration of the homogeneous concrete; flow of mixing water within, or its emergence from, freshly placed or batched concrete.

blockage. When concrete stops moving during concrete pumping: the cause could be mainly due to a faulty mix or equipment deficiency.

blocked filter effect. Effect of fines in the concrete mix which allow the water phase to transmit pressure but at the same time create a high frictional resistance in the pipeline.

blow backs. Air intake into the concrete pump cylinders which sometimes blows out on the piston discharge stroke.

boom. Articulated sections of the pipeline attached to a mobile pump.

boom restriction. Proximity of adjacent structures which affect the full free movement of the mobile pump boom or separate placing boom.

cascade effect. Effect of delays at one site spreading to other sites by reason of ready-mixed concrete delays.

catch basket. Steel basket arrangement attached to the end of a pipeline to retain the wash-out ball and prevent it from being forcibly ejected and possibly causing an accident.

cementitious material. Material in a concrete mix which has cementing properties.

cement replacement. Replacing part of the cement content in a concrete mix with another material, e.g. PFA or GGBFS.

changeover valves. Valves incorporated in a pipeline at a Y junction for changing the direction of the concrete.

chocking, wedging. Fixing the pipeline to a part of a structure using timber pieces or wedges.

cofferdam. Structure usually consisting of steel sheet piling driven to form watertight structures.

cohesiveness. Property of the concrete which enables the aggregate particles and cement paste matrix to remain in contact during mixing, handling and placing.

cold jointing. Joint between a layer of concrete which has already set and a fresh layer. To avoid cold jointing the two layers should set at roughly the same time. This can be done by using a retarder in the first layer.

compressive strength. Measured resistance of concrete to axial loading.

consistency. Degree of stiffness in the freshly mixed concrete which ranges from soft to plastic to stiff; ability of fresh concrete to flow.

construction joint. Surface where two successive pours of concrete meet, normally with a keyway or reinforcing bars across the joint.

cushion effect. Air entrainment in concrete causing a dampening effect when concrete is being pumped.

cruise outlook. Operating plant well within its capabilities and not at maximum performance levels.

deformation (of concrete). Ability of a concrete mix flowing through a pipeline to alter its cross-sectional area without causing segregation.

distribution arm. Equipment used at the placing position to facilitate the placing of the concrete.

dominant plant. Important item of plant which takes precedence in planning, i.e. construction activity must be planned to make best use of item of plant; e.g. tower crane, concrete pump.

dosage. Amount of admixture in a concrete mix, kg/m^3.

drag. Negative resistance to concrete flow in a pipeline caused by friction between the concrete and the walls of the pipeline.

dry process. Sprayed concrete process when water is added to the mix by the gun operator at the gun position.

durability. Long-term performance of the concrete in its hardened state; resistance to weathering, chemical attack, abrasion.

early strength. Strength of concrete developed usually during the first 72 hours.

elephant trunk. Swinging pipe valve system developed by Putzmeister.

equalizing wear. Reversing bends in a pipeline to compensate for wear; placing worn (yet still serviceable) pipes at the end of the pipeline.

extremes (in concrete mix). Too much or too little cement and/or sand in the mix which could inhibit pumping.

fibre reinforcement. Steel, glass or polypropylene fibres 12–50 mm long, used in sprayed concrete.

fines. Materials in the concrete mix below 300 μm in size, i.e. sand and cement including cement replacement material.

fines content. Amount of fine material (cement, sand) in a concrete mix.

flapper valve. Simple type of one piece plate valve which swivels to cover inlet and outlet cylinder openings.

flash over. Electrical arc which can occur between high voltage cables and a concrete pump boom, for example, without the two actually touching.

flash set (sprayed concrete). Rapid development of rigidity in a layer of concrete usually achieved by including an accelerator in the mix.

flexible forms. Textile containers of various shapes and sizes which can take the place of conventional shuttering.

flexible hose. Rubber or rubber/canvas hose used at the end of the boom or pipeline for distribution purposes.

flowability. Degree to which a concrete mix will flow by itself.

flow velocity. Speed of flow of the concrete in a pipeline.

freeform design. Variety of curved shapes, the formation of which is facilitated by the use of sprayed concrete.

grading (of aggregates). Composition of various sizes of aggregates which make up a concrete mix.

grading curve envelope. Parameters of grading curves for a type of mix.

gravity feed. Allowing concrete or grout to flow freely down a pipe under gravitational forces.

grout leakage. Grout emerging from a joint in the pipeline or from the end of a pipeline.

gunite. Term given to sprayed concrete when the maximum size of aggregate is 10 mm.

hardstanding. Area on which a concrete pump or other item of plant is positioned and which should be capable of fully supporting the loads involved.

harsh mixes. Mixes which lack workability and consistency due to insufficient sand or cement or to poor aggregate grading.

high pressure water. Water pumped through a pipeline by compressed air.

hollow transfer tube valve. Type of valve which consists of a tube which alternates between the face of the two cylinders.

hook time. Time a crane is occupied in moving material, i.e. time a load is attached to the hook.

hopper. Container at the rear of the concrete pump which receives the concrete.

hopper grille. Protecting iron bars across the top of the receiving hopper of a concrete pump.

hose end. Point where the concrete emerges at the end of the flexible hose in a pipeline.

hydration. Chemical reaction between water and cement.

hydrovalve. System of placing concrete underwater incorporating the use of a flexible tube.

intangible costs. Costs in an activity of construction to which it is difficult to apportion actual costs at the time.

integrity testing. Testing of concrete piles for strength, cross-sectional area, etc.

kinking. Bending of a flexible hose which could stop the passage of concrete.

leavings. Residual particles of concrete left in a truckmixer after it has discharged its load: may be lumps of old concrete or pieces of large aggregate.

line pressure. Pressure of the concrete in the pipeline at any point.

lubricating film. Cement/water or cement/sand/water mix which is pumped through the pipeline for lubrication purposes.

lubricating the pipeline. Pushing/pumping of grout through the pipeline before starting concrete pumping.

mattresses. Textile mattresses filled with concrete and used for revetment, coastal defence and embankment work.

mechanics of concrete pumping. What actually occurs when concrete is pumped through a pipeline; the basic movement of the concrete.

method statement. Statement drawn up detailing the method to be used for carrying out an operation, e.g. a concrete pour, which should include the responsibilities of all the parties concerned.
microsilica. Minute spheres of silicon dioxide which is a by-product of the manufacture of ferro-silicon: used as an admixture.
mix design. Calculated quantities of the various constituent materials in a concrete mix to give a specified strength and other requirements.
mobile pumps. Concrete pump mounted on a wheeled chassis.
modular principle. Concrete pump manufacturers' method of combining certain sizes of pump with certain sizes of boom or separate placing boom.
mortar pumps. Small pumps for the conveyance of mortar, usually for screeds.

nomograph. Diagram designed by pump manufacturers, based on theoretical outputs, from which a reasonable estimate of the size of pump can be calculated from given requirements, e.g. production rate required, slump, pipeline length.
non-dispersible concrete. Concrete containing an additive which enables the concrete to resist segregation when placed underwater.

one piece shoulder type coupling. Type of pipe coupling constructed in one piece, i.e. not of welded construction.
outriggers. Stabilizing arms of a concrete pump which must be fully extended when pumping.
output (of pump). Placing speed which can be achieved using a specific pump, m^3/h.
oversanded. Mix containing more sand than necessary for adequate workability and satisfactory finishing characteristics.

peristaltic pump. Pump that uses a system of rollers and rubber tube for moving the concrete, sometimes referred to as a squeeze pump.
permeability. Ability of a material to permit the passage of water through it.
pipe cleaning head. Equipment attached to the pipeline for cleaning purposes, which incorporates inlet valves for water and air.
pipe coupling. Steel coupling which joins two lengths of delivery pipe together.
pipeline. Pipe through which the concrete is delivered, constructed in steel generally with a flexible hose at the delivery end.
placing crew. Personnel laying the concrete.
placing rate. Rate at which the concrete is placed in situ (into the formwork), m^3/h.
plastic concrete. Freshly mixed concrete in a cohesive, readily remouldable and workable state.

plasticizer. Admixture which increases the workability of the mix without adding more water.
plastic state (of concrete). Concrete which has not yet set.
plug flow. Movement of pumped concrete in a pipeline.
pneumatic gun. Term given to the spray nozzle equipment used in sprayed concrete work.
pneumatic pump. Pump that uses compressed air to move the concrete.
pozzolanic activity. Cementitious behaviour in the presence of free lime.
pulsations (in pipeline). Movements in the pipeline caused by the action of the pump.
pumpability. Ease with which concrete can be pumped.
pumpable (concrete). Concrete in its plastic state which is capable of being pumped.
pumping pressure. Pressure exerted on the concrete.

quick release couplings. Pipe couplings with raised arms.

rail mounted (pumps). Concrete pumps mounted on a rail chassis for use mainly in tunnel construction.
random dispersion. Adding fibre reinforcement to the mix by hand.
rebound. Materials which rebound off the sprayed surface during sprayed concrete work.
recommended test pressures (of pipeline). Test pressures laid down by manufacturers for various sizes and types of pipeline.
reducing pipe. Tapered pipe for connecting one pipe to another of lesser diameter.
release cock. Valve incorporated in the pipeline for venting pressure.
remote control. Equipment for controlling boom movement and pump operation, used at a distance from the pump: radio or cable system.
resistance to abrasion. Degree to which concrete will resist being worn away by rubbing or friction processes.
retarder. Admixture included in the concrete mix which delays the setting of the concrete.
reverse pumping. Action of sucking back the concrete into the pump, releasing the pressure in the pipeline.
Rock valve. Schwing valve system used in the company's piston pumps.
rotating chutes. Special chutes at the placement position which can be moved radially to assist in placing concrete.

safe working pressure. Maximum pressure at which the concrete pump should be operated, or at which the pipeline should be subjected to.
screeding. Laying sand/mortar material for floor screeds, or

levelling the concrete surface by tamping or mechanical means.

segregation (of concrete). Tendency in flowing concrete for coarse aggregate and drier material to remain behind and for mortar and wetter material to flow ahead, thus causing disintegration of the mix.

self compaction (of concrete). Concrete with properties which enables it to reach a state of compaction without having to use extraneous means.

self levelling. Concrete which is soft and flowing enough to level itself out without the use of levelling equipment.

separate placing boom. Placing booms which are not attached to the pump and placed at strategic positions on site.

shooting arm. Adjustable arm to which the spray gun is attached on mobile sprayed concrete units.

shotcrete. Method of sprayed concrete where the aggregate size is normally larger than 10 mm.

shut-off valves. Valves incorporated in the pipeline to stop the flow of concrete at any time.

site batched. Concrete which has been prepared on site.

skid mounted. Mounting of concrete pumps on to steel channels instead of wheels, used mainly for static pumps.

slewing circle. Arc through which a mobile pump boom or crane jib can move.

sliding gate valve. Early type of valve used in concrete pumps, where a plate was moved alternately across the face of each piston.

slugs (of concrete). Lumps of concrete emerging from the end of the pipeline which look very dry.

sponge ball. Rubber or similar material type ball used for cleaning out a pipeline.

sprayed concrete. Spraying concrete on to a surface.

spray gun. Hand-held gun or nozzle used by the sprayed concrete operator, or mounted on a spraying arm on sprayed concrete units.

stability calculations. Manufacturer's calculations against overturning mobile pumps in all possible positions of the boom when the boom is filled with concrete.

starting distance. Distance from the pump outlet to the first bend.

static pipeline. Steel pipeline for conveying the concrete.

static pump. Concrete pump which will normally be used at one position on site for a considerable time.

static pressure. Pressure created in a vertical pipeline due to the weight of the concrete in the pipeline.

stowing or S bends. Steel pipeline bends used to stop or slow down the flow of concrete when placing downhill.

substrate. Surface on to which sprayed concrete is applied.

tacky concrete. Concrete which is in a state between wet and dry.

test panels (for sprayed concrete). Rigid panels set up at the same angle as the area on which concrete is to be sprayed, from which cores are taken and subsequently tested.

theoretical length of pipeline. Measured length of pipeline plus allowances for bends and tapers, used when calculating the theoretical pressure in a pipeline.

theoretical rating (of pump). Output of a concrete pump, usually given by the manufacturer, m^3/h.

thixotropic (quasi-thixotropic). Property of some gels of becoming fluid when stirred and returning to the original state when stirring ceases.

trailer pump. Concrete pump normally mounted on a single axle which is towed by a vehicle.

tremie. Placing concrete through water using a pipe or tremie tube.

truckmixer concrete pump. Unit consisting of a ready-mixed concrete truck combined with a concrete pump plus placing boom.

tunnel pumps. Pumps which are designed to work in tunnels or placed at tunnel entrances and capable of long distance pumping.

tunnel train. Several units working together for the conveyance of concrete or concreting materials for tunnel construction.

ultrasonic testing. Testing of cast-in-situ piles by electronic means for strength, cross-sectional area and integrity generally.

undersanding. Lack of sand in the mix.

utilization (of plant). Proportion of time during a normal full working day when an item of plant is actually being used, e.g. a crane is moving materials, a pump is pumping concrete.

valve (concrete pump). Mechanism used in a concrete pump for directing the concrete into the delivery line; mechanism controlling the movement of the concrete.

vermiculite plug. Vermiculite particles placed in a tremie pipe to separate the concrete from the water.

void content. Amount of air spaces in a concrete mix.

wall frictional values. Frictional resistance in a pipeline arising from the passage of the concrete along the walls of the pipeline.

wall thickness. Term to describe the thickness of pipes.

wash-out adaptor. Equipment used when washing out a pipeline; has a water and compressed air valve.

wash-out area. Area on site provided by the contractor where the pump or ready-mixed concrete truck can be washed out after completion of work.

wash-out ball. Rubber or sponge ball pumped through the pipeline for cleaning purposes.

water plug. Retaining a plug of water in a pipeline by using two rubber balls, used for cleaning purposes.

wear and tear. Wear on a piece of equipment which occurs from normal use.

wedging. See chocking.

wet process (sprayed concrete). Concrete is already hydrated before it reaches the spraying gun.

whiplash. Sudden movement of a flexible hose, usually following a blockage and subsequent release of pressure: can result from kinking in the flexible hose.

workability (concrete). Property of fresh concrete which determines the ease with which it can be mixed, placed or finished.

work factor (of pump). Factor to be applied for downgrading the theoretical rated output of a concrete pump.

working range (concrete pump). Limits of height, length, depth at which a concrete pump can operate.

worm pumps. Type of pump sometimes referred to as screw-type; pulsation-free type of pump used in grouting and screed pumping.

Abbreviations

ACI	American Concrete Institute
ACPA	American Concrete Pumping Association
ASTM	American Society for Testing Materials
BACMI	British Aggregate Construction Materials Industries
BCA	British Cement Association (formerly the C&CA)
BCPA	British Concrete Pumping Association
BRMCA	British Ready Mixed Concrete Association
BSI	British Standards Institution
C&CA	Cement and Concrete Association (now the BCA)
CAA	Cement Admixtures Association
CFA	Continuous flight auger (piling)
CITB	Construction Industry Training Board
CP	Code of Practice
CS	Concrete Society
ERMCO	European Ready Mixed Concrete Organisation
GGBFS	Ground granulated blast-furnace slag
HMSO	Her Majesty's Stationery Office
HSE	Health and Safety Executive
ICE	Institution of Civil Engineers
OPC	Ordinary Portland cement
PFA	Pulverized-fuel ash
QA	Quality assurance
QSRMC	Quality scheme for ready mixed concrete
SCA	Sprayed Concrete Association
SFRC	Steel fibre reinforced concrete